2 **He** Helium
		5 **B** Boron	6 **C** Carbon	7 **N** Nitrogen	8 **O** Oxygen	9 **F** Fluorine	10 **Ne** Neon
		13 **Al** Aluminium	14 **Si** Silicon	15 **P** Phosphorus	16 **S** Sulfur	17 **Cl** Chlorine	18 **Ar** Argon
28 **Ni** Nickel	29 **Cu** Copper	30 **Zn** Zinc	31 **Ga** Gallium	32 **Ge** Germanium	33 **As** Arsenic	34 **Se** Selenium	35 **Br** Bromine	36 **Kr** Krypton
46 **Pd** Palladium	47 **Ag** Silver	48 **Cd** Cadmium	49 **In** Indium	50 **Sn** Tin	51 **Sb** Antimony	52 **Te** Tellurium	53 **I** Iodine	54 **Xe** Xenon
78 **Pt** Platinum	79 **Au** Gold	80 **Hg** Mercury	81 **Tl** Thallium	82 **Pb** Lead	83 **Bi** Bismuth	84 **Po** Polonium	85 **At** Astatine	86 **Rn** Radon
110 **Ds** Darmstadtium	111 **Rg** Roentgenium	112 **Cn** Copernicium	113 **Nh** Nihonium	114 **Fl** Flerovium	115 **Mc** Moscovium	116 **Lv** Livermorium	117 **Ts** Tennessine	118 **Og** Oganesson

63 **Eu** Europium	64 **Gd** Gadolinium	65 **Tb** Terbium	66 **Dy** Dysprosium	67 **Ho** Holmium	68 **Er** Erbium	69 **Tm** Thulium	70 **Yb** Ytterbium	71 **Lu** Lutetium
95 **Am** Americium	96 **Cm** Curium	97 **Bk** Berkelium	98 **Cf** Californium	99 **Es** Einsteinium	100 **Fm** Fermium	101 **Md** Mendelevium	102 **No** Nobelium	103 **Lr** Lawrencium

The Elements
We Live By

The
Elements
We Live By

How Iron Helps Us Breathe,
Potassium Lets Us See, and Other Surprising
Superpowers of the Periodic Table

ANJA RØYNE

Translated by
OLIVIA LASKY

THE EXPERIMENT

NEW YORK

THE ELEMENTS WE LIVE BY: *How Iron Helps Us Breathe, Potassium Lets Us See, and Other Surprising Superpowers of the Periodic Table*
Copyright © 2018, 2020 by Anja Røyne
Translation © 2020 by The Experiment, LLC

Originally published in Norway as *Menneskets grunnstoffer: Byggeklossene vi og verden er laget av* by Kagge Forlag AS in 2018. First published in North America in revised form by The Experiment, LLC, in 2020.

This translation has been published with the financial support of NORLA.

N NORLA
NORWEGIAN LITERATURE ABROAD

The Experiment, LLC | 220 East 23rd Street, Suite 600
New York, NY 10010-4658
theexperimentpublishing.com

THE EXPERIMENT and its colophon are registered trademarks of The Experiment, LLC. Many of the designations used by manufacturers and sellers to distinguish their products are claimed as trademarks. Where those designations appear in this book and The Experiment was aware of a trademark claim, the designations have been capitalized.

The Experiment's books are available at special discounts when purchased in bulk for premiums and sales promotions as well as for fund-raising or educational use. For details, contact us at info@theexperimentpublishing.com.

Library of Congress Cataloging-in-Publication Data available upon request

ISBN 978-1-61519-645-6
Ebook ISBN 978-1-61519-646-3

Cover and text design by Jack Dunnington
Translated by Olivia Lasky
Author photograph by Kari Margrethe Sabro
Endpaper illustrations courtesy of Line Monrad-Hansen

Manufactured in the United States of America

First printing June 2020
10 9 8 7 6 5 4 3 2 1

Contents

*Introduction: Our Fantastic and Catastrophic
Relationship with the Planet We Live On* 1

**1 | The History of the World and
the Elements in Seven Days** 5

MONDAY: THE BIRTH OF THE UNIVERSE 5

FROM TUESDAY TO THURSDAY: STARS ARE BORN AND DIE . . 9

FRIDAY: OUR SOLAR SYSTEM IS FORMED 9

SATURDAY: LIFE BEGINS 12

SUNDAY: THE LIVING EARTH 16

A HALF SECOND BEFORE MIDNIGHT:
THE AGE OF CIVILIZATION 19

HUMANS AND THE FUTURE 22

2 | All That Glitters Isn't Gold 25

HOW EARTH'S CRUST DID US A FAVOR 26

THE FIRST GOLD 27

GOLD IN THE RIVER GRAVEL 27

THE MINES IN ROȘIA MONTANĂ 29

MINING ON THE SURFACE .32

A TOXIC MEMORY 33

FROM STONE TO METAL 34

GOLD RINGS FROM A TON OF STONE 35

THE END OF ROȘIA MONTANĂ 37

GOLD AND CIVILIZATION .38

THE LOST GOLD 39

3 | The Iron Age Isn't Over 43

THERE'S NO POINT IN BREATHING WITHOUT IRON 44

INTO THE IRON AGE 45

SWEDISH IRON 46

FROM ORE TO METAL 49

COVETABLE STEEL 50

THE PROBLEM WITH RUST 52

CAN WE RUN OUT OF IRON? 54

OUT OF THE IRON AGE? 58

**4 | Copper, Aluminum, and Titanium:
From Light Bulbs to Cyborgs** 61

COPPER IN CARS, BODIES, AND WATER 62

THE COPPER MINES THAT CLEARED THE FORESTS 64

ALUMINUM: RED CLOUDS AND WHITE PINES 66

USING WHAT WE'VE ALREADY USED 69

THE TITANIUM IN A MOUNTAIN 71

THE CYBORGS ARE COMING! 74

THE FUTURE OF MACHINE PEOPLE 77

5 | Calcium and Silicon in Bones and Concrete......81

HARD AND BRITTLE.........................82

MOLDING WITH CLAY.......................83

THE MESSY ATOMS IN THE WINDOWPANE...........84

FROM ALGAE TO CONCRETE....................87

VOLCANIC ASH IN THE COLOSSEUM..............89

CONCRETE THAT SCRAPES THE CLOUDS............91

IS THERE ENOUGH SAND?.....................94

LIVING CERAMICS FACTORIES..................96

6 | Multitalented Carbon: Nails, Rubber, and Plastic...99

NATURAL RUBBER AND VENERABLE VULCANIZATION....100

FROM TIMBER TO TEXTILES....................103

PLASTICS OF THE PAST.......................105

THE TRASH ISLAND.........................107

WHAT DO WE DO WITH ALL THIS PLASTIC?.........109

PLASTIC AFTER OIL.........................111

7 | Potassium, Nitrogen, and Phosphorus:
The Elements That Give Us Food............113

THE JOURNEY TO THE DEAD SEA................114

NUTRIENTS IN OUR NERVES...................116

POTASSIUM FROM WATER.....................118

NITROGEN FROM AIR.......................119

PHOSPHORUS FROM ROCKS...................122

NUTRIENTS GONE ASTRAY...................127

THE FUTURE OF THE DEAD SEA................128

8 | Without Energy, Nothing Happens 129

   ENERGY FROM THE SUN 130

   DRAINING EARTH'S ENERGY STORES 131

   THE SOCIETY WE WANT 132

   ENERGY IN, ENERGY OUT. 133

   OUT OF THE FOSSIL SOCIETY 135

   GEOTHERMAL HEAT AND NUCLEAR POWER:

   ENERGY FROM EARTH'S BEGINNINGS 135

   POWER STRAIGHT FROM THE SUN 137

   WATER THAT RUNS, WIND THAT BLOWS 138

   THE RARE-EARTH ELEMENTS 140

   POWER ON A QUIET WINTER NIGHT. 141

   COBALT IN THE BATTERY. 143

   GASOLINE FROM PLANTS. 146

   TODAY WE EAT OIL 148

9 | Plan B 151

   UNLIMITED ENERGY: A SUN ON EARTH 151

   ELEMENTS IN SPACE 155

   AWAY FROM EARTH? 158

10 | Can We Use Up the Earth? 161

   LIMITS FOR GROWTH 162

   GROWTH THAT IS HAPPENING FASTER
   AND FASTER AND FASTER 164

   THE NECESSITY OF ECONOMIC GROWTH. 165

   CAN THE ECONOMY GROW WITHOUT
   USING MORE RESOURCES?. 168

   AN IMPOSSIBLE PARADOX? 170

   THE HABITABLE ZONE. 170

Acknowledgments 173

References 175

Index 205

About the Author. 214

Introduction
Our Fantastic and Catastrophic Relationship with the Planet We Live On

YOU AND I ARE BOTH a part of the life that once came into existence from the primordial ingredients of our planet. Our bodies are made up of atoms that were formed at the same time as the universe. As my children grow, they're being built from elements in the soil, water, rock, and air. Sometime in the future, the atoms in my body will become trees, glaciers, and granite.

But we humans are also more than just our bodies. The clothes I wear, the house I live in, the knife I use to butter my bread—they're all just as important as my fingers and toes. And without the mines and bulldozers that help make fertilizer and food, you'd probably never even have been born.

All of the objects in our lives—and the materials they're made of—play a role in the unique circumstance we've developed together: our civilization. And I like civilization. I like living in a warm house and traveling to new places, and I can hardly even imagine a life without all the knowledge of the world only a click, tap, or swipe away—even though I grew up with an encyclopedia on the bookshelf and handwritten letters in the mailbox.

Every single day, new windows, phones, and people are being made. It's incredible that this is even possible. But the question is: How do we get hold of the building blocks for all of these people, things, and food? What's it all made of? And will our planet ever run out of these building blocks—causing everything to screech to a halt?

There's a lot of talk about the environment these days, in particular about how human consumption affects water, earth, and air. We talk about species dying out at the same rates as when a meteorite put a hasty end to the dinosaurs sixty-six million years ago. We discuss how the ocean is filled with so much garbage that pretty soon there will be more plastic than fish in the sea. And last but not least, we consider the fact that the oil and coal we're burning in power plants and cars is actually changing the climate—so much so that many places on Earth will become uninhabitable in the near future.

The conversation about environmental degradation can quite easily make me feel completely powerless. Who am I in the big picture, really? Is it my fault that species are dying out? What kind of world am I leaving behind for my children? Is there anything I can do that will not only ease my own conscience but also actually lead to the world evolving in a more positive direction? I wrote this book because I want us to be able to talk about how our steady production of things, food, and, ultimately, ourselves has consequences that are both fantastic and catastrophic at the same time. It's only when we really understand what we're talking about that we can start finding solutions that will actually make a difference for those who come after us.

1 | The History of the World and the Elements in Seven Days

THE HISTORY OF THE ELEMENTS stretches back to the birth of the universe. Their story is long—in fact almost incomprehensibly long in relation to human time. To make it a bit easier, I thought I'd take the creation myth of Genesis as inspiration and tell the history of the world in seven days.

In this story, I'll turn 1 billion years into half a day, 1 million years into 45 seconds, and 1,000 years will be covered in 0.44 seconds. It's been 13.8 billion years since the universe was born, but in this account, time began when the clock struck midnight on Monday morning. By the time you reach the end of this chapter, the clock will strike midnight again, and Sunday will be over.

MONDAY: THE BIRTH OF THE UNIVERSE

At first, there was neither time nor space. How and why everything got started is complicated—but we know that it all started with a bang. The explosion we've come to know as the big bang flung the energy in the newborn universe out in all directions. After this chaotic start, the young universe started to be governed by the laws of nature we know from our world today.

Just as the dust in my house gathers into dust bunnies (it's just a matter of giving it enough time!), the energy in the universe eventually started clumping together. These clumps, or particles of energy, are what we call mass: matter, substance, that which is tangible, that which makes up everything you could potentially touch and feel in the universe.

My body, my belongings, and the planet we live on—absolutely everything we surround ourselves with is made up of atoms. Atoms are composed of three types of particles: protons, neutrons, and electrons. The protons and neutrons are firmly stuck together in the atom's nucleus, and the number of protons in the nucleus is what determines which element the atom is. If the nucleus were to get rid of some protons or receive some new ones, the atom would become a different element. Initially, an atom has the same number of protons and electrons, but the electrons are whirring around the outer edge and can be exchanged between atoms in what we call chemical reactions.

Protons, neutrons, and electrons arose in the glowing soup of energy and mass that made up the young universe. Protons and neutrons ended up sticking together and becoming atomic nuclei in the elements hydrogen, helium, and lithium. These smallest and lightest elements have one, two, and three proteins in their nuclei, respectively. Today, hydrogen is an important building block in water and in the organic molecules that make up living beings. Take the human body, for example, which is made up of almost 10 percent hydrogen. When you think of it that way, you could technically say that you come directly from the birth of the universe!

Sixteen seconds past midnight, the universe had gotten so

cold that electrons could attach to atomic nuclei without being released immediately. So, for the very first time, it was possible for light to move through the universe without being stopped by hot electrons. At just past midnight, there was visible light in the universe—even though there was no one there to see it.

Over the course of the next twelve hours, the mass in the universe continued clumping together. Huge clouds of atoms were formed, and before the clock struck three in the morning, groups of these clouds had become the very first galaxies. One of these galaxies turned out to be the Milky Way—our home. Today, the Milky Way is just one of more than two thousand billion galaxies in the universe.

At 6:00 AM, some of the atom clouds in the galaxies had become so big that they collapsed beneath their own weight. This is how the first stars came to be. In one of these—a clump of material that was considerably larger than our own sun is today—were the hydrogen atoms that would be transformed into the oxygen you just inhaled.

The weight of all the surrounding atoms pressed these hydrogen atoms against each other with enormous force. First, this caused the electrons to detach from the nuclei. The pressure then became so intense that it caused the hydrogen nuclei to fuse together and form new helium nuclei. This fusion released huge amounts of energy that warmed up the clump of atoms, making it a bright star. The same process is still taking place today in our own sun; the light that meets your eyes when you look outside your window comes from atomic nuclei fusing within the sun's interior.

As most of the hydrogen nuclei gradually became helium, the release of energy in the star's interior started to slow down. The center of the star no longer had enough power to withstand the pressure from the surrounding material, and it collapsed. This started a new phase of the star's life. The collapse forced the helium nuclei so close together that they fused in new reactions. Three helium nuclei with two protons each became a nucleus with six protons—which is carbon. Then, the carbon nucleus fused with yet another helium nucleus to form a nucleus of eight protons. This is oxygen, and this atomic nucleus can be found at this very moment in an oxygen atom inside a red blood cell on its way to your brain.

Inside the star, the process of fusing atomic nuclei into heavier and heavier elements continued. Eighty-six percent of your body is made up of carbon, nitrogen, and oxygen, all of which were formed during this phase. Here on Earth, the pressure is far too low to make such elements, so we can be sure that these building blocks in our bodies actually did come from stars. We are stellar beings—every single one of us! In addition, the iron in our blood, the phosphate in our skeletons and DNA, the aluminum in our mobile phones, and the salt (sodium and chloride) we sprinkle on our food were all made during this phase.

A few minutes into the weeklong story, the star's life is over, ending with an explosion so spectacular that it got the name "supernova." In the explosion, elements even heavier than iron were formed—including nickel, copper, and zinc. The power lines in your house are made of materials from a supernova.

The leftovers from the explosion—the material that was not thrown into space, that is—collapsed and became a neutron star. In a neutron star, all of the nuclei have fused together into a massive clump the size of a large city (about 10 miles/15 kilometers in diameter), and, in a way, it really is an enormous nucleus, even though we don't call it an element. There are about one billion neutron stars here in our own galaxy, but since they're so small and cold compared to the other stars, it's not easy to spot them.

When I think about how much space there is in the universe and how small neutron stars are, I feel like what happened next seems almost infinitely impossible. All the same, we know that it had to have happened. At some point during the first days of the universe, two neutron stars collided. This collision created gold, silver, platinum, uranium, and a host of other elements so heavy that they can only be formed in such extreme events. The newborn elements were cast out into space and mixed with clouds of dust and atoms in the galaxy.

And that's how elements came about on the first of seven days. Elements are still being created out in the universe as stars are being born and dying, exploding, and colliding all the time. Here on Earth, however, the elements are fairly constant. It's only through radioactive processes in which unstable nuclei of uranium and other heavy elements sometimes start splitting up that elements are created and destroyed on our planet. Even in laboratories, it's almost impossible to recreate the processes that take place inside stars. We have almost endless opportunities to create materials by varying how we assemble elements, but when it comes to the elements themselves—what we've got is what we've got.

FROM TUESDAY TO THURSDAY: STARS ARE BORN AND DIE

The universe continued on the same track for the next three days. Stars were born, and stars died. Supernovae sent pressure waves and clouds of matter out into space. Since hydrogen and helium were constantly being fused into new elements inside the stars, the total amount of hydrogen and helium in the universe steadily decreased while the amount of heavier elements increased.

FRIDAY: OUR SOLAR SYSTEM IS FORMED

At four o'clock on Friday afternoon, a star died in our neighborhood. The pressure wave from the supernova squeezed dust and gas into a cloud that contained the oxygen you just inhaled. This triggered a chain reaction where clumps of matter became heavy enough to absorb the dust and gas in the surrounding area, and the bigger and heavier they became, the more of their surroundings they sucked up. Just forty-five minutes later, the cloud had become a star with several planets in its orbit. This star is our sun—the center of our solar system.

All planets orbit around a star. The closer to the star the planet is, the more the planet is heated by the radiation from the nuclear reactions in the star's interior. In our solar system, the closest planets became extremely hot. Today, they have surface temperatures of over 750°F (400°C). The outer planets, on the other hand, are quite cold; the sun's rays can't get them any warmer than 32°F (0°C). The planets farthest away are frozen worlds below –300° F (–185°C).

But for one planet, the distance from the sun was just right. In the habitable zone around the sun, the temperature of the planet could be low enough that water doesn't boil and at the same time high enough that not all water freezes. It was this planet that would become our home, Earth.

In the beginning, however, Earth was glowing hot—fully liquid, actually. It was also constantly being hit by large and small meteorites. One or more of these stones hit Earth with such force that the matter thrown off after the collision clumped together in orbit around Earth, becoming the moon.

As Earth gradually cooled off in the cold of outer space, heavy elements such as iron, gold, and uranium sank into the center of the liquid sphere. The lighter elements—including silicon and the main components of our bodies, carbon, oxygen, hydrogen, and nitrogen—were left at the outermost edge, eventually forming a solid crust of siliceous rock around the planet with a gaseous atmosphere surrounding it.

In this first atmosphere, molecules were formed—groups of atoms in which two hydrogen atoms were linked to one oxygen atom—this is water. At 6:30 in the evening, the temperature had become cool enough for water molecules to clump into droplets. When the droplets had become large and heavy enough, they rained down on the surface, creating the first, warm sea.

Deep down in this sea, something almost magical happened: Carbon, hydrogen, and oxygen attached themselves to large molecules along with smaller amounts of sulfur, nitrogen, and phosphorus. At some point, some of these molecules developed a structure that caused them to make copies of themselves by getting nearby elements to attach in the very same way. This is the basis of life. When did these molecules go from being a complex chemical system to becoming something living? Did life arise in one place, at one time, or did life first spread across the planet after a long series of attempts? Researchers do not yet have a clear answer for this, but we ourselves are proof that life succeeded.

We humans will never benefit from the metals that sank into the middle of the planet; they're simply much too far into the center of the earth. Fortunately, something happened around ten o'clock on Friday night that would be decisive in how we've been able to develop our society: For the rest of the evening, Earth was bombarded by meteorites. Scientists don't quite understand why. One theory is that the larger planets were adjusting their orbits and disrupting how other matter was moving in the solar system. In any case, the metal in these meteorites was cast out across Earth's crust without sinking into the center because the crust had become firmer. These are the metals we use to make cars and forks today.

About half an hour before midnight, Earth's crust started to crack open and move. The crust on our planet still consists of plates floating around the mantle, a viscous sea of rock. Up here on the surface, it's so cold that when molten rock appears through cracks between the plates, it solidifies and becomes new crust. The plates are therefore constantly changing shape as they move in relation to one another. When the continents on two different plates collide, large mountain ranges are formed—just as the Himalayas are being formed right now as India squeezes into Asia from the south. In many places, a plate with a thin seabed slides under the thicker continental crust of another plate. This is happening today along the Pacific coast of South America. Elsewhere, the plates scrape against one another, shoulder to shoulder. If they get stuck, huge earthquakes can be triggered when they finally slip again, crushing bedrock and leaving large systems of cracks throughout the bedrock.

The dance of Earth's plates with one another is known as plate tectonics. In our solar system, Earth is the only planet with such an active surface. It isn't clear why only Earth's crust is dancing, but without this dance, it would've been a dead planet. Plate tectonics is Earth's conveyor belt and the driving force behind everything that makes our planet such an exciting place. The movement recycles Earth's materials by allowing what has been carried out to sea by water and wind and then buried in the seafloor for millions of years to be lifted back up to the surface. It creates cracks through the crust where flowing water can transport elements up from the depths. Today, the remnants of these cracks are where we excavate gold and other metals.

SATURDAY: LIFE BEGINS

The bombardment of Earth's crust lasted until about a quarter to one early Saturday morning. Then, conditions on the planet became a lot calmer. By 5:30 in the morning, Earth had developed its own magnetic field—an invisible shield that prevents most of the sun's energy-rich and harmful particles from reaching the planet. Without this protection, we would have to live in subterranean caves in order to survive.

Around the same time as the formation of the magnetic field, the first single-celled organisms appeared.

In reality, living organisms aren't anything but small machines that use energy from their surroundings to make copies of themselves. The organisms can, of course, have several other functions as well, such as registering what's going on around them, moving, or communicating with one another. While our bodies get energy from the food we eat, researchers believe the very first living creatures harvested the energy they needed from chemical compounds deep in the oceans. There are still entire ecosystems that live in complete darkness in areas where the tectonic plates are sliding apart. Here, mineral-rich water flows up through chimneylike structures in the seabed, and the chemical bonds in these minerals contain energy that living beings can benefit from.

Today, almost all life on Earth gets its energy from the sun, either directly through photosynthesis or by eating molecules that contain stored solar energy. During photosynthesis, the energy coming from sunlight is used to split carbon dioxide and water into carbon, hydrogen, and oxygen. The atoms are then put together in new combinations to form the energy-rich molecules we know as carbohydrates, proteins, and fats. Photosynthesis was most likely developed by bacteria in the ocean around three o'clock on Saturday afternoon. Today, it's still being utilized in all green plants, trees, and blue-green algae (cyanobacteria). All matter in these organisms contains a small amount of solar energy.

When carbon dioxide and water become plant material, there is an excess of oxygen atoms. Organisms involved in photosynthesis release these oxygen atoms in the form of oxygen molecules, where two oxygen atoms are bonded to each other. Oxygen molecules have a tendency to react with other compounds. We're familiar with this from fire, which is nothing more than oxygen reacting with carbon or other combustible substances and releasing energy in the form of heat. We therefore wouldn't be able to find oxygen molecules in either the ocean or the atmosphere, if they weren't being produced by some source at any given time. The oxygen gas that is so vital to us is constantly being produced through photosynthesis—but in the very first atmosphere on Earth, there were no oxygen molecules, and none of the first organisms needed oxygen to survive, either.

Before photosynthesis began, the oceans contained large amounts of dissolved iron, which they no longer do. In our day and age, iron that comes into contact with water very quickly develops a rough, red surface that breaks down easily. This red material—rust—is a chemical link between iron and oxygen. As long as there is oxygen in the air and water, unprotected iron will always rust.

On Saturday afternoon between around 3:00 and 6:45, the seas started to rust. All of the oxygen that was produced in the first photosynthesis reacted with iron, and the resulting rust sank to the bottom. Eventually, this rust became thick layers of reddish, banded rock. Today, we dig up this red rock, remove the oxygen from the iron in large furnaces, and use the resulting iron metal to make knives and train tracks.

When most of the iron had rusted away, oxygen molecules started to accumulate in the oceans. Oxygen was a deadly poison for most of the planet's first creatures—and photosynthesis thus led to one of the largest mass extinctions of species our planet has ever undergone. However, there were some creatures that had learned to use oxygen to their advantage, for example by utilizing the oxygen from their surroundings to release solar energy that was stored in organisms they'd eaten. By doing so, they got the energy to run their own life processes without having to perform photosynthesis on their own.

While myriad life-forms were lost to the toxic oxygen, the organisms that did use oxygen gained a tremendous advantage. We are the descendants of these organisms. The energy you're using to read this—to move your eyes and transform the text into information in your brain—comes from a chemical reaction in

which oxygen and carbohydrates turn into carbon dioxide and water inside your body's cells.

As seawater became saturated with oxygen, oxygen gas started to flow from the oceans into the atmosphere. This change led to tremendous upheavals here on Earth. Our planet is constantly radiating heat into outer space, and the temperature of its surface is heavily influenced by how much of this thermal radiation is trapped by gases in the atmosphere. This is what we call the greenhouse effect. The early atmosphere was rich in methane, which absorbs a great deal of thermal radiation, keeping Earth's surface warm. As oxygen gas in the atmosphere started to break down the methane, the greenhouse effect became weaker, and the planet was propelled into a global ice age. Up until a quarter past nine on Saturday evening, much of the biodiversity that had developed in the seas had been lost to the cold.

High up in the atmosphere, oxygen molecules were being struck by the most energy-rich light from the sun, causing the two atoms in the oxygen molecules to break apart. When the single oxygen atoms collided with the other oxygen molecules that were flying past, ozone (molecules with three oxygen atoms) was formed. The ozone layer effectively acts as a trap for the most energy-rich part of the sun's rays, which can torch vulnerable organic molecules if they reach Earth's surface. Today, the ozone layer is what makes it possible for us to walk around beneath the open sky without experiencing severe damage to our eyes and skin.

Once the ozone layer was in place, it was then possible for organisms to survive near the surface of the water, and even on dry land. Here, there was even more sunlight to utilize in photosynthesis, and the production of organic matter and oxygen gas saw a sharp increase. The first forms of life on dry land were mats of algae and bacteria that covered a flat, barren landscape and laid the foundation for what would become a layer of fertile soil on our planet.

SUNDAY: THE LIVING EARTH

The organisms with cell nuclei (which are what we originate from) arose at about 3:20 AM on Sunday. By five o'clock that morning, single-celled organisms had developed such close cooperation that they were no longer considered to be isolated individuals but rather living beings consisting of multiple cells. Nevertheless, it still took a long time before life as we know it really started to flourish. It wasn't until 5:25 PM, after Earth had undergone a new global ice age between 3:15 and 4:15, that specialized species of plants and animals emerged to form complex ecosystems in the oceans. When geologists study petrified seabeds from the subsequent time period, they find fossils from a variety of species, such as cephalopods and woodlouse-like trilobites.

At five past six on Sunday evening, the first animals crawled ashore, where they started working on converting algae-matter and stone into a layer of fertile soil. This is where the first land plants could take root, which they did at about 6:31 PM. With plant roots holding on to soil, water, and later also tree trunks, preventing the wind from blowing away loose materials from the ground, dry land went from being smooth and barren to becoming more diverse—with rivers, valleys, marshes, and lakes.

Life on Earth suffered a few hard blows as well—volcanic eruptions, meteor showers, and changes in solar activity led to major changes in temperature, sea level, and oxygen levels in the atmosphere and in the ocean. Eighty-five percent of the species that had developed after the first flourishing of complex life was lost in a global ice age by 6:36 PM. Life picked up again thereafter, but at 7:28 PM, the trilobites were suffocated by a lack of oxygen on the seabed and disappeared along with 80 percent of all species found in the oceans at that time.

The largest mass extinction to date occurred at 8:56 PM on Sunday evening, when enormous volcanic eruptions in Siberia sent considerable amounts of carbon dioxide into the atmosphere, something that ultimately led to a global temperature increase and acidification of the world's seas—issues we are well aware of today. The fossils from just after the extinction testify to a disaster that left desolate landscapes with neither forests on land nor coral reefs in the seas.

A few minutes later, however, the forests and seas flourished once again, and more and more species came into existence. Both mammals and dinosaurs made their appearance before 9:30 that evening. The good times didn't last long, however, coming to an end at 9:34, when a new global warming wiped out at least three-quarters of all species on Earth. Mammals and dinosaurs were among those who endured, and it was perhaps precisely this extermination of competitors that gave dinosaurs the chance to become the next rulers of Earth. When the dinosaurs also had to give in around 11:12 PM, Earth's climate had likely been so punishing for such a long time that when the tremendous meteorite hit in what is present-day Mexico, it was simply the last nail in the coffin for many of Earth's species.

When there was no longer the danger of being eaten by dinosaurs, mammals could disperse and utilize a wide variety of ecological niches. Initially, the climate was warmer than it is today, but at about 11:25 PM, the temperature began to fall. Eighteen minutes later, many of the planet's lush, dark jungles were replaced with grassy plains. It was these species of grass in particular that were to lay the foundations for human agriculture closer to midnight. At this point, we're starting to approach the age of humankind. Some mammals had already evolved into what we know as apes, and at 11:45 PM, Hominoidea—the branch to which gorillas, chimpanzees, and humans belong—distinguished themselves from the other apes.

Humans separated themselves from the Hominoidea five minutes before midnight—which is where we find ourselves now. It's only been two minutes since our ancestors used the first stone tools to break open animal bones to get at the nutritious bone marrow.

One minute and twenty seconds ago, the globe got considerably cooler, and the planet entered a cycle of ice ages and interglacial periods that have continued into our time. It was therefore crucial for early humans to learn to master fire. However, it seems it wasn't until about thirteen seconds to midnight that bonfires were used in daily life.

Fire kept humans warm, protected them from predators, and enabled them to see each other and their surroundings after the sun had set. By cooking food over fires, people could use the solar energy stored in wood to break down their food rather than needing their own jaws and digestive systems to do all the work. This freed up time and energy that humans could use for other activities, and may have been crucial in our ability to think and communicate.

Our species, Homo sapiens, originated in Africa nine seconds ago. For a long time, we were just one of several human species. We also know of the Neanderthals, who lived in Europe and the Middle East before the emergence of Homo sapiens. The Neanderthals lived side by side with our species until one and a half seconds ago, when they were either outcompeted or simply wiped out by our ancestors. It's only in the last half-second that Homo sapiens have been the only human species on Earth.

As we close in on one second before midnight, Homo sapiens developed language and, with it, the ability to tell each other stories, plan the future, and trade between different peoples. Aided by new technologies such as the bow and arrow, needle and thread, fishing hooks, boats, and oil lamps, they journeyed beyond Africa and occupied the rest of the world.

The first humans lived in nomadic tribes. The members of each group worked together on hunting and gathering wild, edible plants, but also on caring for those who were too young, too old, or too sick to contribute to the community. As we approach midnight with great strides, society becomes more and more recognizable to us modern humans.

A HALF SECOND BEFORE MIDNIGHT: THE AGE OF CIVILIZATION

It's difficult to get a proper feeling for periods of less than one second, so we're going to adjust the scale here. Let's look at the last half-second of the history of the world like a five-hundred-meter sprint in which we've just reached the finish line today. Each meter of the race equals one thousandth of a second, or twenty-three years in real time. If we'd used the same scale throughout the history of the universe, the distance would have amounted to over 375,000 miles (600,000 km)—or as far as the distance from Earth to the moon and almost all the way back again. Our five-hundred-meter sprint starts 11,500 years ago, when humans first started living in the same place for extended periods of time.

When humans transitioned from living as nomads to becoming settled, they were—for the very first time—able to own more than they could carry with them to the next settlement. It then became more relevant to develop specialized tools that were as effective as possible for each purpose. People could also become more specialized; instead of everyone contributing to everything, certain people could spend time on what they were best at. This could be useful for the group as a whole in that some people dedicated themselves to making clothing or tools, while others hunted or gathered plants.

Agriculture probably emerged as a by-product of humans settling in one place: The gatherers returned home with the plants they liked best. The waste from the gathering and cooking contained seeds from these plants, which enjoyed good conditions to germinate near the settlement. This made it easier for inhabitants to harvest from these plants. As humans mostly harvested the specimens most suitable to them—for example, those with the largest seeds—these species gradually changed and became the first agricultural plants. In time, humans found they could improve the crops near the settlement through clearing the land, watering, and plowing. Three hundred and fifty meters before the finish line, humans had become farmers.

Agriculture provided people with a stable and predictable source of food, which enabled the population to grow. However, there were also some disadvantages. Farming is difficult and often monotonous work, and the farmers probably had less free time than the nomads who were their ancestors. They were also on a more uniform diet, which could lead to malnutrition, and they were subject to famine if the crops failed.

It is conceivable that humans lived healthier and perhaps even happier lives before they started farming. However, it was ultimately the cultivation of land that laid the foundations for what we call civilization. The growing and storage of food made a more hierarchical and specialized organization of society both possible and necessary. Two hundred meters before the finish line, people were organized in the first kingdoms and had developed written language, money, and religion. By domesticating animals such as bulls and horses, people also had a new source of physical strength that allowed them to plow and cultivate larger areas, thus providing for even more people. With the help of livestock, they could also travel across greater distances over a shorter amount of time, allowing them to trade both goods and knowledge more effectively than before.

In these specialized societies, humans developed the advanced technologies required for mining and utilizing metals. People produced bronze tools 200 meters before the finish line, and iron ones at 140 meters to go; 88 meters before the finish (around the time of the birth of Jesus), steel had come into production. At 22 meters, mankind went through what has been called the scientific revolution, when new and systematic methods for understanding the laws of nature were developed.

Up until this point, all human activity had in some way or another been driven by the sunlight that reached Earth each day. Solar energy stored in plant material was burned to produce heat or eaten by animals and humans to be utilized as muscle power. In addition, mills could be run by hydropower, thus harnessing the energy of the water that the sun had elevated from the sea to higher altitudes. Sailboats got their momentum from the wind that blew due to the temperature differences the sun created when it shined on Earth.

Eleven meters before the finish, humans started to actively exploit fossil energy sources—that is, solar energy that has been stored in the ground for millions of years. When the daily solar radiation was supplemented first primarily by coal, and then by oil and gas, almost any type of industry could be operated without the risk of running out of fuel due to deforestation of the surrounding areas. The industrial revolution transformed the world of humankind.

The development of antibiotics three meters before the finish provided us with a health-care system that could cure illness in a society in which child mortality was no longer a matter of course and where childbirth didn't put women's lives at risk.

Two meters before the finish, humans traveled into outer space.

Then the clock struck midnight—and we're back to today. We live in a world where countless possibilities lie ahead of us. With our unique capabilities, we seem to be able to overcome any problem we face.

HUMANS AND THE FUTURE

When humans first began transitioning from a nomadic to a more settled lifestyle, there weren't many of us on Earth—maybe even just a couple million in total. During the transition to an agricultural society, the number had risen to 10 million. From then on, the population continued to rise steadily. By the time people started using bronze and iron, the population had increased tenfold to 100 million. Since then, Earth's population has doubled several times. The population just before the birth of Christ was 200 million. In the thirteenth century, when stave churches were being built in Norway and the Mongols were making their way to eastern Europe, the world's population was 400 million. A full 800 million was reached by the industrial revolution in the late eighteenth century. By the end of the nineteenth century, the population had doubled again to 1.6 billion; by the 1960s we topped 3.2 billion; and in 2005 we'd reached 6.4 billion. If population growth continues at today's pace, the next doubling to 12.8 billion people will have occurred by 2068. However, current trends suggest that the population will either stabilize or start to decline before we reach 11 billion. As I write this, the world's population has passed 7.7 billion people.

In another thousand years, we'll have gone forty more meters. This is negligible in the greater scheme of things, but still longer than we tend to think about when we're planning for the future. As Earth's population increases, we use more and more of the planet's resources. Will we always have enough elements to make ourselves and all of the things we need? Will we have the energy required to extract elements from Earth's crust? Will people in a thousand years be able to look back on as fantastic a development as the one we've had thus far?

Humans' ability to extract and use metals is unique among all animals on Earth, and it all started with gold. We associate gold with power, wealth, and adventure, and such stories are themselves among the most important building blocks of our civilization. Let's start there.

2 | All That Glitters Isn't Gold

I'VE WORN A GOLD RING for over ten years. Everyone who sees it knows what it means; the wedding ring is a well-recognized symbol of love and commitment.

A few years before I exchanged rings with my husband, the two of us spent a summer traveling around Europe by train. We started with the night train to Copenhagen before traveling east through Germany, the Czech Republic, Slovakia, and Hungary to Romania. From there, we made our way to the area called Transylvania, a place we'd previously only known as Dracula's home.

Transylvania was like another world. We didn't find any vampires, but we did see park rangers with scythes on the bus. Horse and carriage were as common on the roads as cars. Poor and run-down areas stood side by side with faded historic buildings that testified to a rich—but for me, unknown—past.

Had I read about the area beforehand, I would have known that Transylvania's wealth was built on gold. Then I would have been able to visit the place that houses Europe's largest known gold deposit: the old city of Roșia Montană, which is now in danger of being buried beneath its own mining waste. Wealth from the earth never comes free.

HOW EARTH'S CRUST DID US A FAVOR

Twelve million years ago, what is present-day Transylvania was full of active volcanoes. Hot, molten rock pushed up through Earth's crust before breaking through the surface, creating clouds of ash and lava flows. Deep beneath the surface, the bedrock was warmed up by the magma, and as a result, the water that had been trapped in crystals within the stone began to be released. All of this water started to seep upward through the solid rock—first as tiny droplets, and then as small, slow streams through the crevices and voids the magma had opened in the crust.

It wasn't just water that was released, though. The rock the water flowed through also contained tiny amounts of gold. Usually, water has no effect on gold; for example, gold bars in sunken pirate ships can stay sparkling for hundreds of years. However, in these extreme environments where the water was several hundred degrees and contained large amounts of chlorine and sulfur, even the very resilient gold had to yield. One by one, the gold atoms attached themselves to sulfur atoms and were then transported upward along with the water.

As the water flowed up, the pressure went down. Just like when steam comes out of the open valve of a pressure cooker, somewhere along the way the pressure became so low that the water started to boil. Then, the sulfur atoms "let go" of the gold to instead bind themselves to the water vapor. The abandoned gold atoms scurried back to one another and were left behind as layers of shiny metal.

Over the course of thousands and millions of years, large gold deposits were built up beneath the Romanian volcanoes.

Then one day, the volcanic activity came to a halt. Earth's crust cooled off, and for millions of years, the surface of the rock was worn down by weather, water, and wind. Valleys, hills, and mountains were formed and changed.

And then came the people.

THE FIRST GOLD

Maybe it happened ten thousand years ago: A child was playing in a little river on a warm, sunny day. Suddenly, she saw the sun glistening on a very special rock on the riverbed. She lifted it up, amazed at how heavy it was. As she pounded it against another stone, she saw, to her amazement, that there was a mark in her shiny new plaything. This stone did not resemble anything she'd played with before.

After the other children saw what the little girl had found, it didn't take long before they found several more golden lumps in the riverbed. The adults, who started to investigate the new material, found that they could hammer and mold it into thin layers and intricate shapes. The beautiful objects aroused the interest of their neighbors, who were keen to trade it in exchange for other goods. As more people had their eyes opened to gold, they discovered more and more places where it could be picked up by the riverbeds. This could be how metal became a part of human life.

GOLD IN THE RIVER GRAVEL

No matter how it really began, researchers believe that gold was the first metal that was mined and utilized by people. Although gold is rare, it's remarkably easy to find and use in comparison to other metals, mainly because it is found in nature in its metallic form.

Like all other elements, gold has a specific number of protons in its atomic nucleus. This number is what determines how elements behave since they control what relationship the element has to the electrons buzzing around the nucleus. Any chemical reaction involves some form of exchange of electrons between atoms. Some elements are desperate to get rid of one or more electrons, while others are constantly searching for extra ones they can borrow from others. Gold, however, is satisfied as it is and therefore thrives with other gold atoms and is prone to form a pure metal. This also means gold rarely participates in chemical reactions: It's not particularly interesting for the machinery that is our bodies. We adorn ourselves with gold, but the gold we might find in our bodies are minuscule particles that have ended up there by mistake.

Since gold occurs in metal form in nature, it's possible to simply pick it up off the ground—or, more often, from the bottom of a river. Gold finds its way to the riverbeds when rocks that contain veins of gold are broken down. This allows clumps of gold to come loose and be carried down the river with other rocks and gravel. Incidentally, it's actually incorrect to talk about "clumps" of gold; although there are indeed sometimes large clumps, most of the gold in rivers and mountains are tiny particles mixed with another stone. Since gold is so heavy, the grains can separate from the rest of the gravel with the help of gravity—using a pan and the right technique. This is likely how the first large-scale extraction of gold took off, perhaps not very far from Roșia Montană. The gold that nature had flushed out of the Roșia Montană deposits could have been extracted in an organized fashion as early as five thousand years before our time.

According to Greek legend, the hero Jason was challenged by his uncle Pelias—who'd stolen the throne from Jason's father—to bring home the Golden Fleece from a land far away. If Jason came back with the treasure, he would be able to take over the throne. The journey took Jason to an area by the Black Sea, where he got hold of the Golden Fleece after defeating the dragon that guarded it. There are a number of interpretations about what the fleece in this story could symbolize—royal power, say, or the introduction of sheep husbandry. Recently, however, researchers have come to realize that it might quite literally symbolize a golden sheepskin. It turns out that over three thousand years ago, both in Egypt and the area around the Black Sea, sheepskin was used to sort tiny particles of gold from rock dust. The surface of the gold particles differs from the surface of most other minerals in the stone, which can cause the gold to adhere to certain substances—such as the water-repellent surfaces found in sheep's wool. Gold dust can therefore be collected by allowing a mixture of water and gravel to flow over a sheepskin, attaching to the hairs against the stream. This method may have been used for a long time but was forgotten after the fall of the Roman Empire.

Today, there aren't that many places where you can find gold in the river gravel. Most of it has already been collected by people seeking fortune. When we've found all of the gold nature itself has broken out, we'll have to go straight to the source: the gold-bearing rocks.

THE MINES IN ROȘIA MONTANĂ

Before I started learning about geology, most of my knowledge about gold mining came from cartoons, which had me thinking that gold was broken loose from veins of pure metal. Unfortunately, this isn't actually the case. When a stone contains enough of a metal where it would be worthwhile to extract it, it's called ore. Gold ore most often contains much of the white or transparent mineral quartz, with tiny grains of gold spread throughout. If wind and weather haven't ground the rock to pieces, it's up to humans themselves to break the ore to get to the gold. This is hard work, since the atoms that build up the rock are linked with incredibly strong bonds. It wasn't before people developed proper iron tools that they could really start extracting gold from solid rock.

The first gold miners in Transylvania used fire setting to make the job easier. This technique involves lighting a fire against the side of the rock and letting it burn until the stone becomes glowing hot. Stone expands when heated up, and the minerals in the stone expand in different directions. This creates a network of small and large cracks in the stone, which makes it easier to break loose. Fire setting was probably already used by the Dacians, the people who lived in Transylvania before the Roman invasion. Fire setting was also used in the mines in Norway up until the end of the nineteenth century. However, while fire setting is a useful technique, it does have some drawbacks. For one, a single bonfire takes you no more than a few inches into the rock. Huge amounts of firewood were required to operate a mine, which led to a great deal of pressure on the forests in the area. The fires also made the air difficult to breathe for workers in deep mines.

The Romans were the mining masters of antiquity. The Dacians likely recruited Roman engineers to help them build up mining operations in Transylvania. However, this may have cost more than it was worth, as it ultimately opened the Romans' eyes to the rich gold deposits in Transylvania—and in the year 106 CE, they conquered the kingdom of Dacia. One of the reasons historians believe the mines were already being run before the Roman invasion is that the Romans took such large quantities of gold from the Dacians—as much as 165 metric tons. It seems unlikely the Dacians could have extracted all of this solely by means of panning in the rivers. After the invasion, the Romans established the city of Alburnus Maior, later called Roșia Montană, and sent their best gold mining specialists there along with thousands of engineers and slaves. Over the course of just fifty years, they'd built up one of the largest mining complexes in the Roman Empire.

The gold from Alburnus Maior ended up being a tremendous source of wealth for the Romans and financed huge expansions in Rome. However, it still wasn't enough to prevent the fall of the Roman Empire, and the Romans had left the area as early as 271 ce. By then, they'd managed to make miles-long tunnels in the mountains, about 4 miles (6 km) of which are still preserved today. With the fall of the Roman Empire, much of the Romans' highly developed mining technology was lost, although mining operations in Roșia Montană did continue in simpler forms.

It wasn't until the end of the eighteenth century that major advances were made in Roșia Montană's mining operations. The Habsburgs, who had been in control of Transylvania over a long period of time, developed water-powered crushing mills that were supplied by artificial dams farther up in the mountains. Previously, miners had to use more or less manual methods to crush the loose rock into dust, but with several hundred newly built crushing mills, the gold mines in Roșia Montană once again became a source of great wealth, and the activity helped the city flourish. Miners came from all parts of the Habsburg Empire, and churches, bars, banks, and casinos were established—several of which still exist to this day.

In 1867, the Habsburg Empire became the Austro-Hungarian Empire, which crumbled after World War I. Transylvania became part of Romania, and the mines were distributed to private individuals. Mining continued to flourish in private hands until the Communists took over in 1948 and nationalized all industry, including the mines in Roșia Montană.

MINING ON THE SURFACE

After millennia of increasingly intensive mining, the richest gold deposits in the rock beneath Roşia Montană were nearly depleted. As in all mining operations, the most accessible is taken first. The Roman tunnels followed the veins where the gold was most concentrated. As there was gradually less and less gold and more quartz in each of the stones transported from the mines, it became more and more expensive to mine the gold. When the mountain had become riddled with nearly 90 miles (145 km) of passages, mining here was no longer profitable—which is why the Communists switched from underground mining to open-pit (quarry) mining in the 1970s.

Rather than digging tunnels down or into the mountainside, open-pit mining involves simply removing everything that covers the ore you're interested in. You then work your way downward into an enormous pit. The development of large and heavy machinery—not to mention superior and more precise explosives—has made this an economically viable solution for many mineral deposits. While you do have to move larger amounts of rock in a quarry than in an underground mine, the transport itself is cheaper and easier when the stone can be loaded into large trucks than when it needs to be transported out of deep mining shafts, where you also have to worry about stability, ventilation, and drainage.

Today, society is focusing more on reducing the environmental impacts of mining than was the case a few decades ago. These days, a quarry often gets started by transporting the topsoil that needs to be moved to a suitable storage location. Then, the mining company usually has to remove a portion of rock that does not contain ore. This is also stored somewhere, such that when the operation in one part of the quarry is completed, it can be filled with stone and soil once again. Over time, new vegetation will do its best to conceal the gaping wound in the terrain.

A TOXIC MEMORY

However, the pit that remains after a quarry isn't the only trace a mining operation leaves in the landscape. Gold still needs to be separated from tailings (leftover ore waste), and to do this, the gold ore must be pulverized and mixed with water. Then, the same principles of the good old days are used: washing pans and sheepskins, albeit on a considerably larger scale. The mixture of rock dust and water is first poured into a chute designed to catch the heaviest gold particles. Next, large machines are used to mix in a soaplike additive and stir vigorously while supplying air, causing the mixture to foam. Just as gold dust gets stuck in sheep's wool, it also adheres to the soap bubbles. This dust can be scraped from the surface of the tub, while the uninteresting minerals remain in the water and sink to the bottom as sludge.

This sludge can't be used for anything, but it does take up space. In Norway and some other places, this type of sludge has been deposited underwater, in a fjord (I'll get back to this later). Most often, the mining company will build a dam at the mouth of a valley such that the valley can be used as a kind of "sludge pool." Not far from Roșia Montană is the village of Geamana, where four hundred families were forced out of their homes when the state decided that the valley the village was located in would be used as a disposal site for the Roșia Poieni copper mine. Today, the once lush valley has become a muddy plain where nothing grows, with extraterrestrial patterns in rust-red and green patchworking across the wasteland. The only sign of the town's existence is the roof and spire of Geamana's church, which is still sticking out of the mud on the plain.

Stone is rarely toxic in itself, but large quantities of crushed stone can still create serious environmental problems. Most types of rocks contain minerals that can react with water. These reactions occur very slowly within intact rock, as the minerals are so tightly packed that water never reaches then. When the stone is pulverized, however, water gets everywhere. Rainwater or ground-water will almost inevitably seep through mine waste dumps, and when the water comes out of these landfills, it will have reacted with the rock dust and absorbed heavy metals that can damage downstream ecosystems. From a human perspective, this process will continue eternally. The very first mines in history are still producing pollution that damages the ecosystems around them.

FROM STONE TO METAL

The gold nuggets and dust sorted out from the sludge and tailings still aren't clean enough to make gold jewelry. Before gold can be converted into bars and sold on the international market, it must first be separated from the rock dust it's lodged in. Mercury was often used for this purpose in the past. Mercury is a toxic metal that is liquid at room temperature and has the unique ability to dissolve gold. In a mixture of mercury and gold ore dust, a blend of mercury and gold forms in which all impurities are collected on the surface of the liquid metal—such that they can be scraped away. In the end, the gold can be separated from the mercury by heating the metal mixture until the mercury is steamed off.

Today, most mining companies have made the transition to utilizing a safer—but still toxic—material: cyanide. Cyanide is a compound of carbon and nitrogen found in many places in nature (such as cherry pits), and in small amounts it can quickly break down into other harmless materials. For most of us, cyanide is probably best known as an ingredient in hydrocyanic acid, which was used in Nazi gas chambers during World War II. When gold ore dust is mixed with cyanide-containing water, the gold dissolves. Other chemicals are then added that cause the rest of the dust to clump together and sink to the bottom. Finally, the water is mixed with fine zinc dust, which absorbs the cyanide. The gold atoms can then find their way back to each other and become particles of gold metal.

Cyanide is also used to extract gold directly from unsorted, crushed gold ore. It's a tremendous advantage to be able to avoid using energy to pulverize the stone, so this method makes it profitable to extract gold even in small concentrations. The ore is heaped in large mounds atop a dense layer of plastic or clay. At the top of this mound, a network of pipes with small holes is made—and then the process of irrigating the mound with water containing cyanide begins. This water seeps through the mound and into a large collection of ponds. From the air, these ponds are a lovely turquoise but will hopefully always have a net stretched across them to prevent birds from landing and dying in the toxic water.

In the year 2000, Romania was hit by what has been called Europe's largest environmental catastrophe since the Chernobyl incident. The dam that was holding the cyanide-containing water from a gold extraction project in Baia Mare, close to the Hungarian border, broke. The water ran into the Someş River, which flows into Hungary's second largest river, the Tisza, and then on into the Danube. The accident contaminated the drinking water of millions of people and killed almost all life-forms in certain parts of the rivers—but thankfully took few or no human lives. A public investigation held no company responsible. Despite such spectacular examples of accidents and pollution, cyanidation is still considered to be a relatively safe way of extracting gold and is currently used by over 90 percent of the more than five hundred gold mines across the globe.

GOLD RINGS FROM A TON OF STONE

As almost 1,700 metric tons of gold have been mined, there are still well over 300 metric tons left in the rock beneath Roşia Montană. However, the total amount of gold in a deposit alone isn't enough to determine whether a deposit is worth mining. Equally important is how concentrated the gold is. The concentration indicates how much stone must be dug out and processed in order to obtain this gold. When a mine is established, a mining company first extracts from the areas where the ore has the highest concentration of metal—since this is where there's the most money to be made. Thereafter, the operation moves to ore of a lower and lower concentration until it is no longer economically profitable. The next 100 metric tons of gold that will be extracted from Roșia Montană will present greater challenges than the first 100.

Let's take my own wedding ring as an example. It is smooth, two millimeters wide, weighs five grams, and is made of fourteen-carat gold. A carat is a historical unit of measurement; if you were to separate all the metals in the ring from each other and place them in twenty-four piles of equal weight, the number of carats would tell you how many of the piles were gold. Since pure gold is too soft to be made into functional jewelry, wedding rings are most commonly fourteen-carat (or 58 percent) gold. In other words, I walk around with about three grams of gold on my finger, while the rest of the ring is made up of copper and silver.

Today, the average concentration of gold in the ore mined around the world is between one and three grams of gold per metric ton of stone. So if the gold in my ring was mined recently, it would mean it came from an entire metric ton of stone. This is equivalent to a boulder you and two good friends could sit on together: about 1.5 feet (0.5 m) wide, 1.5 feet tall, and 4 feet (1 m) long. To make my ring, this boulder had to be blown up, crushed, pulverized, processed, moved, and dumped into a refuse pit. A hundred fifty years ago, on the other hand, the concentration in the mines would have been between 0.75 and 1 ounce (20 and

30 g), so the same boulder would have provided enough gold for ten wedding rings instead of just one.

The more stone that has to be processed for every gram of extracted gold, the more energy, chemicals, and space the mining requires—and this costs money. Although Roşia Montană is still Europe's largest (discovered) gold deposit, after more than two thousand years of extraction, the failing economy forced mining operations in Roşia Montană to an end in 2006.

THE END OF ROŞIA MONTANĂ

Today, a fight over Roşia Montană is underway, a fight between international mining companies and national environmental interests. However, there's also a clash between the inhabitants who want new jobs and the activity that accompanies new mining, and those who prefer to continue working in agriculture and tourism in today's landscape.

In order to extract the remainder of the gold from the rock, a new project proposes to open four new quarries operated with cyanide extraction. What's new in this proposal is that, in order to get the last remnants of the gold, the city of Roşia Montană must be buried. The city that was founded on gold could end up being wiped out by the very activity and resources that have helped it flourish for centuries. The remains of the Roman mines (that weren't already destroyed by the Communists) would be wiped out as well. Local opponents are working to get Roşia Montană placed on the UNESCO World Heritage list.

Should the project be approved, 250 million metric tons of waste from the cyanide extraction process would be dumped in the valley. Four churches would be buried, and the mining companies have already begun compensating people who wish to dig up their dead relatives from the six cemeteries that would also be wiped away—so that they can bury them elsewhere.

The mining company argues that they're focusing on the environment, and that they'll devote considerable resources to cleaning up the pollution from mining days of yore.

GOLD AND CIVILIZATION

Is gold important? Is it worth all this destruction?

Gold is one of the least abundant elements found in Earth's crust but is nonetheless found in extractable deposits in most parts of the world. Gold is quite easy to identify due to its color and weight, making it difficult to cheat by replacing gold with any other material. Objects made of gold can also stay shiny for centuries. These properties have made gold an outstanding means of payment and a symbol of wealth. As a universally accepted currency, gold has been crucial for trade between nations and for the development of civilization as we know it.

It's still the case in politically or economically unstable countries that it may seem safest to have some of your assets in the form of gold. It was only after I'd spoken about this with friends from the Middle East—who told me how important it is to give gold as wedding gifts and for other special occasions—that I realized that traditions of baptismal jewelry and silverware are also rooted in something more than their beauty. Gold provides financial security.

With two major world events occurring in 2016—namely when the British voted to leave the EU and when it became clear that Donald Trump would actually become the forty-fifth president of the United States—the price of gold rose significantly. As a result, the production of gold items in the United States was about the same in 2016 as the year before, but a greater proportion of this was as gold bars and coins procured by nervous investors, while less jewelry was made and sold because of the high price of gold.

Most of the gold used to produce new things today still goes to jewelry and coins, but more than a third is now also used for electronics. Gold is a good electrical conductor, and since it doesn't rust or form surface coatings that could hamper conductivity (as with most other metals), it is sought after for electronic circuits. Gold can also be used in nanometer-thin layers of glass to control how light is reflected off the surface. Advanced transmissions of light can in many cases take over the role electricity has had to date, for example in our transition from copper lines to fiber networks for the internet in our homes. All the mobile phones and computers we own contain gold as well.

THE LOST GOLD

Since we prize gold so much, it is perhaps the element for which we have the best estimate of the deposits that exist in the world. Today, few believe that many new, large gold deposits will be found in the future. The total amount of gold that can be extracted from deposits in Earth's crust is estimated to be about 333,000 metric tons. Of this, we've already extracted 187,000 metric tons and made it into jewelry, coins, and other objects. This means there's something like 146,000 metric tons of gold that remain to be extracted, and that mankind is over halfway into its history as gold miners. Just as in the case of Roșia Montană, the most recently extracted gold was more difficult to access than the first. We first took large gold nuggets from the rivers. Then we chopped out the gold ore that sparkled the most. In order to get the last tons of gold, we'll have to work to a greater extent on the rocks that house it.

According to detailed inventories of all of the gold in bank vaults, warehouses, and jewelry boxes, the amount of gold in human possession in some form or another totals about 181,000 metric tons. This is 6,000 metric tons less than what's been extracted throughout history. So what happened to the rest?

Unbelievable as it may sound, it is estimated that several hundred tons of gold are lying in sunken vessels on the seafloor. In addition, around a thousand tons of gold are buried in cemeteries in the deceased's gold teeth and jewelry. Perhaps grave robbery will emerge as a macabre source of gold in the future. . . .

A few thousand tons of gold are found in waste, in the form of discarded electronics and machinery. The gold in electronic components is mixed with a variety of other substances. Recycling is therefore not as simple as collecting the gold and recasting it. As the amount of gold in electronic waste increases while it simultaneously becomes more expensive to extract from geological deposits, more and more gold mining will have to come from human sources. Today, this is called "urban mining." In the same way as sophisticated methods for separating gold from tailings have been developed over time, researchers are now in the process of developing chemical and mechanical processes best suited to separate gold from plastics and other metals in mobile phones.

In 2016, 3,100 metric tons of gold were extracted from the world's mines. At today's rate of extraction, they'll be empty within less than fifty years. In a thousand years, we'll have to expect that all available gold has been transferred from rock and stone into the technosphere, which is what we call everything humans have created and use.

Both the gold in ships on the seafloor and in teeth in cemeteries could also be collected and reused. The gold that has truly been lost to mankind currently amounts to around a thousand tons. This is gold that has been placed as a surface coating on various objects—for example, gilded furniture and picture frames—and worn off with time. Some of the gold in electronics could also fall into this category when electronic waste ends up in the wrong place. This gold is lost. It's become dust and been blown or washed away by wind and water. It will end up at the bottom of the ocean and become a part of the sediment along with everything else that's been washed out by rivers. In a few million years, it will become stone. And then—after many millions of years—volcanic activity in this area may start to produce warm, sulfuric water that flushes through the rock, dissolves the gold, and transports it upward until it settles in cracks and voids and builds up into gold ore over a long, long time.

3 | The Iron Age Isn't Over

I'M LYING STILL, BUT MY bed is moving. Tomorrow, I'll be attending a meeting on the other side of the mountains. Tonight, I'm being rocked to sleep by the movements of a train, just as generations of people have been before me. These kinds of train trips make me feel like I'm part of a long story; the railway has transported people and goods through landscapes and between cities since long before we ever had cars and planes.

Both the trains and railway tracks are made of iron—the most important metal in our civilization. The first metals mankind started using—gold, copper, and bronze, which is an alloy of copper and tin—are in most cases too soft to replace tools made of stone and wood. The use of iron, on the other hand, led to a revolution in both warfare and agriculture. Just imagine the difference between plowing a field with a plow made of wood versus one made of iron. Iron tools made it easier to cultivate land, build roads, and chop wood. In combination with iron weapons such as arrowheads and swords, this metal provided a tremendous advantage to those who mastered it earlier than their neighbors.

THERE'S NO POINT IN BREATHING WITHOUT IRON

But iron doesn't just play an important role in our society; it also constitutes a crucial part of the body's own transportation system. An adult human body contains about four grams of iron atoms (enough for a medium-sized nail), and we use the iron in our bodies to perform a vital task.

I need to breathe in order to live. All the cells in my body require oxygen. When I take a breath, I get oxygen in my lungs, but I still need a way of transporting that oxygen further to my cells. This is where iron comes in. Unlike gold, which prefers to hang on to its own electrons, iron is a generous creature that is more than happy to give some away. This creates a close friendship between iron and oxygen, as oxygen is always eager to take extra electrons from other elements.

When blood meets the air in my lungs, oxygen takes the opportunity to attach itself to iron atoms that are bound to molecules in my blood. In this way, blood transports oxygen farther into the body. There are other molecules in my cells that persuade iron and oxygen to part ways again, and the solitary iron continues through the blood vessels all the way to the heart, where it is then pumped back into the lungs to fetch new oxygen. Without iron in my blood, it doesn't matter how much I breathe, since I wouldn't actually be able to use any of the oxygen I so laboriously inhale into my lungs. This is why I have to take iron tablets for a few weeks after donating a pint of blood to the blood bank. My body can easily create new blood cells on its own—but not iron.

Once iron has emitted electrons to oxygen, it requires a lot of energy to separate the two elements from one another again. It took a long time for people to learn how to break this bond and give the iron atoms their electrons back, which is something that must be done to turn iron into a metal that can be made into weapons and tools.

INTO THE IRON AGE

When the Egyptian pharaoh Tutankhamen was buried 3,500 years ago, he had an iron dagger with him in his sarcophagus. This, as well as other ancient iron objects that have been found across the globe, remained a great mystery for a long time; after all, the methods needed to produce the iron metal these objects were made of wasn't developed until about a thousand years later.

The explanation to the mystery lies beyond our planet: The metal in Tutankhamen's dagger doesn't come from Earth.

In outer space, there are quantities of small and large asteroids made up of nickel-containing iron. Here, they are exposed to neither water nor oxygen—which means that the iron in asteroids doesn't rust and can stay shiny and metallic forever. Every once in a while, some of these asteroids tumble down to us on Earth in the form of meteorites that can simply be picked up off the ground and hammered into daggers and other objects. This was the very first iron metal we humans used. All such early iron objects were most likely made from meteorite iron.

There are very few places on Earth where natural iron atoms aren't bound to other elements but rather appear in metal form. One of these deposits is somewhere in Greenland, where ferrous lava penetrated Earth's crust long, long ago. On its way up, this molten rock pushed through a layer of coal—the remnants of prehistoric plants consisting almost exclusively of carbon. One of carbon's most useful features is that it's even more eager to emit electrons than iron is. As a result, when the bonded iron and oxygen atoms in the glowing hot magma came into contact with carbon in the coal, the carbon managed to "persuade" the iron atoms to take on their extra electrons. Carbon and oxygen poured into the atmosphere in the form of carbon dioxide, and a layer of iron was left behind, ready to be used by us humans.

Here lies the key to the production of metallic iron—a key that humans had to find before they could enter the Iron Age. We have plenty of iron around us (Earth's crust contains about 4 percent iron), but almost all of that iron is bound to oxygen. Iron ore can be converted into iron metal by being mixed with coal and heated up until the coal catches fire. Then, the carbon in the burning coal reacts with the iron ore, emitting electrons and "stealing" oxygen, therefore leaving the iron behind in metal form.

When humans started to produce iron, it also led to an increased need for timber. When timber is heated in closed pits without oxygen, it becomes coal that can be used in iron production. This often led to extensive strain on local forests, with deforestation occurring as a common and unfortunate side effect. Today, we use fossil coal that we dig from the ground to produce iron, thus eliminating the need to chop down trees. In a way, coal mines have played a considerable role in saving much of the world's forests from ending up in coal pits. At the same time, though, the fossil carbon that is released into the atmosphere when we burn fossil coal contributes to heating up our planet. For every ton of iron produced, about half a ton of carbon dioxide is also created from the carbon from the coal and the oxygen from the iron ore. In the long run, this may pose an even greater threat to forests and ecosystems than the previous logging ever did.

SWEDISH IRON

While the forests of the past provided us with the coal we needed for iron production, iron ore itself is the result of organisms that lived even longer ago. Almost all of the iron ore we dig up today originates from the rust-red layers of iron oxide that developed on the seafloor when photosynthesis first began and the seas rusted around 2.5 billion years ago. Today, these deposits can be found as horizontal layers close to the surface and are therefore well suited for extraction in quarries. The earth and stone above the iron ore are dug up and set off to the side so that the ore can be broken out with enormous machines from huge, bowl-shaped pits in the terrain— some of the largest human-made structures here on Earth.

Since iron is such a common element, there are also deposits of iron ore that were created by other means. One of the most important is in the town of Kiruna in the far north of Sweden. Both this town and the railway that leads there were built to mine iron out of the northern Swedish rock.

For some time, people had been aware of the rich iron ore deposits in the area where Kiruna is today, yet this part of northern Sweden was almost completely deserted up until the end of the 1800s. The high levels of phosphorus in the iron ore made it almost worthless on the world market, but when a method was developed to remove phosphorus from iron ore, this Swedish iron became a sought-after raw material.

The deposit in Kiruna was remote, and it could take several days to transport the ore via reindeer and sleds to the port in Luleå, deep in the Gulf of Bothnia, the branch of the Baltic between Finland and Sweden. During wintertime, the ice was often so thick that the ore had to stay piled on land until open water made it possible to transport it to the rest of Europe. In the spring of 1898, the Swedish Parliament therefore decided to build a railway that would link Kiruna to both Luleå and the Norwegian port of Narvik over 100 miles (160 km) away. This considerable investment would make it possible to deliver iron ore to the world year-round. The development attracted thousands of Swedes, Norwegians, and Finns keen on earning money from mining, railway construction, or any of the other professions that accompany such activities, including skilled trades, alcohol sales, and prostitution. After a rather unremarkable start, Kiruna quickly developed into a proper city with schools, hospitals, and fire stations.

The railway was completed in 1902, and with it, Kiruna was established as an important source of iron ore for all of Europe. One of the largest customers was Germany, and at the start of World War II, Hitler was entirely dependent on this supply. More than half of the iron he needed to produce tanks, bombers, and weapons for the German army came from Kiruna. The supply lines were secured when Germany occupied both Norway and Denmark on April 9, 1940, and shipments from Kiruna to Germany continued until they were stopped by the Allies in 1944.

The iron in Kiruna originates from molten rock that penetrated Earth's crust sometime in the past. As magma slowly cooled in the cavity it had created inside rock, crystals of ferrous minerals were formed that sank to the bottom of the magma chamber, which is how iron was separated from the rest of the elements in the magma. Today, the bottom of the old magma chamber slopes steeply downward in the rock, which is what makes Kiruna one of the few iron mines in the world in which ore is excavated underground. Deep down in the rock, large tunnels are drilled into the layers of ore before the rock is blasted loose from the bottom up. This causes the rock to tumble from the cave ceiling to the ground, where it is crushed by the fall, collected in trucks, and transported up to the surface. Here, the ferrous minerals are sorted out and loaded onto trams.

When the rock is dug out and tumbles down into the depths, there will inevitably be cracks that extend upward toward the surface. Today, the cracks beneath Kiruna have reached so far up that the downtown area will soon sink into the honeycombed rock. The city can no longer remain where it is. The church and some other select historic buildings are now being put on wheels and moved to firmer ground, where they will be surrounded by newly built schools, shops, and homes for all of the people who will soon need to pack their moving boxes.

FROM ORE TO METAL

Trains with iron ore from Kiruna still arrive in Narvik several times a day, every single day. In Narvik, the ore is loaded onto ships to be transported to ironworks around the globe. Today, China is the world's largest producer of iron metal, followed by Japan and India.

In ironworks, the ore is heated up with coal in gigantic furnaces. The coal emits electrons to the iron and takes away oxygen. As the temperature rises in the furnace, the minerals that are not desired in the ultimate product start to melt. This liquid mass is called slag, which can be poured or scraped away from the iron ore. At the end of this process, the ore has become a sticky, spongy lump of "pig iron" that still contains a lot of carbon from the coal.

In the past, this was the type of iron that was used to make objects. The pig iron was hammered to get rid of most of the slag remnants. Then, the blacksmith could heat up the iron until it was red hot and use a hammer and anvil to forge weapons and tools. In Scandinavia, the Vikings produced such chunks of pig iron in specially constructed furnaces on farms, most often with iron ore extracted from nearby marshes. The blacksmith had to know how to control the temperature, air supply, and hammering of the iron in order to get the best possible result.

It was later discovered that the metal's quality could be improved even further if it was melted down one more time. Iron that contains a great deal of carbon and other impurities stays liquid at a temperature low enough that the iron can be poured into molds. This is called cast iron—the cheapest form of iron metal produced today. We find it in our kitchen pots and pans, as well as in a large number of machinery parts.

Wrought iron, which we know from black decorative fences and chandeliers, is produced by fusing pig iron with lime and other substances that help get as many of the impurities out of the metal and into the slag as possible. As the iron becomes cleaner, the melting temperature rises. When the iron can no longer stay liquid in the furnace, it can be taken out, hammered, and forged. The Eiffel Tower—which was completed in 1889—is made of wrought iron.

COVETABLE STEEL

The most sought-after form of iron metal has its own name and has become the very symbol of strength. It does indeed sound impressive to have "arms of steel" or "nerves of steel." Steel is an iron metal with very low levels of carbon—no more than about 1 percent. Steel production was incredibly expensive until the end of the nineteenth century, when new technological developments made it possible to produce the metal on a large scale. Prior to this, however, it was reserved for only the most important objects—such as swords and elastic steel springs.

Even though the amount of carbon in steel is minimal, steel is still technically an alloy of iron and carbon. An alloy is a metal consisting of a mixture of two or more elements and can have properties completely different from those of the individual elements it's made of. It isn't like mixing sugar and salt and getting something that tastes both salty and sweet. Strong steel is made of iron, which in its pure form is soft and bendable (and therefore not particularly well suited for tools), and carbon, which we know from the crumbly graphite found in pencils. A number of other elements can also be added to steel to give it special properties. Small amounts of metals such as vanadium and molybdenum make steel both lighter and stronger and can be found in wrenches and many of the other tools we find in our garages. Chromium creates a steel that doesn't rust as easily and is included, together with nickel and manganese, in the stainless steel cutlery I eat my dinner with.

In order to understand why a specific material behaves as it does, we need to know how the atoms in the material are arranged. If you took a piece of pure iron metal and looked at it under a high-magnification microscope, you'd see that the metal consists of many small, interconnected crystals without any space between them. Unfortunately, you can't see single atoms with ordinary microscopes, but if you could, you would see that in each crystal, the iron atoms are arranged in neat rows.

If you try to bend a pure iron rod with your hands, a row of atoms can quite easily slip past the next. As soon as you stop applying force to it, the atoms stop in their new positions and remain there. The iron rod won't bounce back to its original shape when you release it, as a steel spring might. The amount of force you need to apply to bend the rod is determined by the size of the crystals. Where the crystals hit each other, the rows of atoms are at different angles, which hinders their sliding movement; that's why it's easier to bend an iron rod with large crystals than a rod with small ones.

In the liquid metal produced in the furnace, carbon and iron atoms are well mixed. When this melt cools down, crystals of pure iron start to form. The iron is separated out from the liquid metal while the carbon isn't, thus increasing the proportion of carbon in the remaining melt. This continues until the temperature in the furnace has become so low that not even the remaining mixture of iron and carbon can stay liquid. A new substance is then formed: iron carbide, which consists of one-quarter carbon atoms and three-quarters iron atoms. The gaps between the iron crystals are filled with layer upon layer of iron carbide and pure iron. The final, solid metal consists of a mixture of pure iron crystals, which are elastic, and the layered carbonaceous material, which is stiff.

This combination of stiffness and elasticity is what makes steel so valuable. Thanks to its elasticity, an overloaded steel bridge won't collapse without warning. Instead, it will just bend slightly—and still be nearly as strong afterward. One of the most important applications for steel today is as reinforcement in concrete structures. Concrete can bear a great deal of weight but cracks easily if bent or overextended. When concrete is reinforced with steel, the steel rods inside are able to withstand the forces that bend or extend the structure while the concrete bears the heavy load that would have caused the steel rods alone to bend and give out.

THE PROBLEM WITH RUST

Steel solves a lot of our problems—but not once and for all. The use of iron is a perpetual struggle against the forces of nature. When we make iron metal from iron ore, we use a great deal of energy, forcing the iron atoms into a state they don't really want to be in.

We've all seen that the shiny metal on our car hoods or bicycle frames gets stained by a porous red material after some time. This is rust, the result of iron atoms transferring electrons back to oxygen again, and, unfortunately for us, it's iron's "preferred" form here on Earth's surface. Society therefore spends a lot of money and energy on counteracting rust or corrosion, and on repairing the damage that occurs when corrosion is inevitable.

Although iron and oxygen would more than willingly exchange electrons with each other, they need water for this reaction to take place. The first, and most often simplest, defense against rust is therefore to protect the iron surface from coming into contact with water. The Eiffel Tower, which is made of wrought iron, is painted so all the surfaces get a new coat every seven years. This is how the tower has been kept in such good shape, despite having been built well over a hundred years ago.

Painting is an easy fix, but it isn't always practical. Nobody wants paint on the cutlery they eat their dinners with; it could peel off and mix with the food. Instead, we use stainless steel—an alloy of iron and chromium—where a reaction between the steel and the oxygen in the air causes a dense membrane of impermeable material to form on the metal. This membrane prevents oxygen from further reacting with the iron. Ordinary steel also rusts on the surface, but in this case, the rust forms a porous layer that falls off easily in large flakes, such that the reaction isn't prevented from continuing inward.

It's much more expensive to produce stainless steel than regular steel, which is why stainless steel isn't used to build large structures such as ships, bridges, or oil platforms. Metal structures that are completely or partially submerged in water would also be impossible to protect with the help of paint since it would wear off so quickly. Pieces of zinc or magnesium are often placed on the steel hulls of ships so that these metals will rust instead of the iron. Such metal pieces are often called sacrificial anodes. They function as long as they consist of metals that are even more eager to emit electrons than iron is.

Sometimes, it's easiest to just accept that things will rust. Steel poles that can't be painted have to be made extra thick so that they won't fail even if the surface rusts away. An estimated four millimeters will rust away in moist soil over a hundred years—or as much as thirty millimeters in a hundred years in seawater or areas exposed to lots of sea spray.

Our infrastructure was built with loss in mind. Rusty iron is washed away by rainfall. Paint gets worn down, washed, or blown away. Sacrificial anodes of zinc, aluminum, and magnesium dissolve and disappear into the oceans. We also lose iron when it's worn down; a dull knife edge must be sharpened, and then a thin layer of material is removed. The sharp teeth on bike sprockets also get rounded down with use. The material that was in the teeth becomes dust on the roadside, and with time it will wash out into rivers and eventually the sea.

Still, steel objects are made to last—and they do. Stainless steel cutlery can last for at least 100 years, and bridges, railroad tracks, and skyscrapers can last for between 50 and 150 years before requiring significant structural maintenance. We therefore have a large, growing stock of iron in our society that can be recycled and reused in new structures.

CAN WE RUN OUT OF IRON?

Iron is the world's cheapest and most commonly used metal. In 2016, 1.64 billion metric tons of steel were produced globally—twenty-two times more than what we produced of the next most common metal, aluminum. Over the course of the last 170 years, iron production has increased by between 5 and 10 percent each year. We use iron and steel to make buildings, bridges, railway tracks, ships, trains, buses, cars, high-voltage pylons, and hydropower plants. Iron is the most important ingredient in the most important parts of our infrastructure. We are all people of the Iron Age.

Can you imagine if we ran out of iron? The consequences would be catastrophic. In some cases, we could, of course, replace iron with other materials. Other metals often do work better than iron when we need something that is lighter, like aluminum, or that conducts electricity better, like copper, or that can be put inside human bodies without rusting, like titanium. In other cases, we can replace iron with nonmetallic materials. A bridge can be built from wood instead of steel. Boats can be made from fiberglass and plastic. Knives can be made of ceramic materials. However, even if we were able to produce as much as we could from everything else, we wouldn't even come close to being able to replace all the iron with other materials while at the same time maintaining the society we have today.

It's difficult to say for sure how much of different elements can be extracted in the future. The most reliable numbers we have of this are called reserves. These are based on mining companies' public estimates of what they'll be able to extract from their mines.

Sometimes the news will report that we only have five years left of one element or twenty years left of another. We get these figures by adding up all of the documented reserves of a particular element and dividing this sum by how much is extracted each year at current levels. The resulting number of years will then tell us how long it will take before we've used up the reserves of a given element. The published iron reserves are now at 83 billion metric tons, and 1.5 billion are extracted from mines each year. In other words, if we continue extracting just as much as we do today, we'll have used up the reserves within twenty-eight years. And if we continue to increase production, they'll be empty even sooner.

If this were actually true, we'd been in a pretty bad spot—but luckily it isn't. In fact, this "lifetime" of the iron reserves has been fairly constant for a long time. Fifty years ago, we had only a few more decades left of reserves, and the same goes for other metals; between 1980 and 2011, we had only thirty years of copper and sixty years of nickel—despite the fact that production of both metals doubled over this period.

The reason is quite simple: The reserves tell us how much the mining companies know with certainty that they'll be able to extract from a given area. They don't indicate anything about the deposits we haven't found yet. Because the reserves are a part of the valuation for mining companies, they must undergo a costly process of geological exploration, test drilling, approvals, and certifications in order for deposits to be classified as reserves. It's important for mining companies to document enough reserves to secure the investments required to start or continue mining, but they don't really need any more than that. So, it doesn't make any sense to document reserves for centuries of future use, even though it should be obvious that they exist.

If technological development leads to a new, greater area of application for an element or war breaks out in one of the central production countries, the reserves can become small in relation to the production—and estimated life expectancy shrinks. The element becomes scarce, and prices increase. At the prospect of higher prices, mining companies will choose to use more resources on finding and classifying new reserves. What appears to be a potential halt in supplies can therefore actually lead to new discoveries and increased reserves.

As prices rise, deposits that are already known can also be moved into the reserves category. This is because reserves only denote the deposits that can be extracted for financial gain. When prices are higher, mining companies can afford to dig deeper, crush more stone, and use more expensive and sophisticated sorting methods.

Technological advances can also create new reserves. The iron ore from Kiruna was long considered to be unusable due to its high phosphorus content, but a new method for removing phosphorus from pig iron took Kiruna from being a wasteland to holding a key position in European politics. In the future, robots' entry into mining operations will allow them to dig deeper and sort more efficiently, which is one way that reserves will continue to grow in the years to come.

It is in the nature of the reserves that they will grow when we need them to. Some use this as an argument to say that we'll never really experience a shortage of resources; we can always find more, or develop methods of extracting more. This might not really be the case, though. Everything moved into this little box called "reserves" already existed in the box called "known resources." This box contains all of the deposits we've already found but have not yet classified as reserves since they're too far off the beaten track, or are in a country at war or that doesn't allow mining out of environmental consideration, or because the geology is such that extraction isn't profitable in today's market or with today's technology.

The final box holds the unknown resources—everything we haven't yet discovered because we haven't mapped out every cubic foot of Earth's crust.

Each time we discover a new deposit, it moves from the "unknown" box to the "known" box. It can then be moved further into the reserves box. Each time the reserves grow, the resources shrink correspondingly. They may be unknown, but they aren't infinite, and when the resource box is empty, there will be nothing left to extract.

Those who've tried to estimate the total resources for iron have ended up somewhere between 230 and 360 billion metric tons. In addition, between 30 and 70 billion metric tons have already been extracted and are either somewhere in our society or lost to rust and wear.

We've been using iron for over 3,000 years but have only extracted perhaps a tenth of all of the iron that's available. This doesn't mean that we can continue as we have been for another 10,000 more years, though; most of the iron we've mined has come from the last century. The ratio between today's iron production and the total resources indicates about 250 more years of production.

Since the numbers for unknown resources are so uncertain, we could very well assume that they are in fact four times larger, and that we would therefore have enough iron for a thousand more years. Or maybe they're even ten times larger. Then in a thousand years, we wouldn't be at the end of the Iron Age, but perhaps only halfway through. The problem is that resource scarcity doesn't happen when Earth's crust is empty; we'll have a shortage of iron as soon as our society is no longer able to afford to extract it.

OUT OF THE IRON AGE?

How long will we be able to afford iron? This is a difficult and important question. If the number of people on Earth continues to increase, iron consumption will most likely rise as well. If more people earn more money, it will also lead to increased demand. Population decline and hard times could lead to less need. New technology can create new markets or eliminate what used to be major markets—thereby driving demand up or down.

Of all of the known and unknown resources, only some deposits are of good quality. Here, the iron concentration is high, meaning there's less need to blast, move, crush, sort, and store large quantities of stone to get the iron we need. When the best deposits are used up, we'll be forced to get started on the ones with lower concentrations. The consequence is that for every ton of iron we extract from the planet's stocks, we'll be forced to spend more energy and money to extract the next ton. This can lead to iron becoming more expensive, which would make it more difficult for some of us to buy iron tools. It could also lead to the infrastructure we're so dependent on becoming more expensive to build and maintain.

Quite recently, researchers have looked at all of these mechanisms in context in order to glean more information about what development we can expect for the extraction and use of iron over the next four hundred years. They started by assuming that the total iron resources are 340 billion metric tons, and that the world's population will have reduced sharply by the year 2400 (this is connected to the prospects they found for other elements, including phosphorus, which we'll get back to later). According to the results of their studies, iron extraction from mines will continue to increase into the middle of this century and then decline, as it will become increasingly more expensive and energy intensive. The rising price of iron ore will also affect the price of scrap iron, meaning that more iron will be recycled. By the end of this century, most of the iron on the market will be produced from scrap, while these days it's under a third. Well into the twenty-third century, mining will have virtually ceased. At the same time, corrosion and wear will cause iron loss to continue as before. We can't reduce these losses by making more stainless steel since there will also be shortages of the alloys in stainless steel—namely chromium, manganese, and nickel—long before iron becomes scarce. The amount of iron being used and owned by humans will increase from about 50 billion metric tons today to almost 160 billion metric tons by mid-2100. When this scenario ends in the year 2400, humans will be left with only about 30 billion metric tons of iron.

We should never consider a single study to be the full truth, especially when we're trying to predict what will happen so far in the future—but assuming that nothing lasts forever, it seems reasonable that our descendants will have to take the first steps out of the Iron Age.

Iron is vital to our civilization. For Hitler, the iron from Kiruna was so important that he occupied Norway and Denmark to secure the supplies. It isn't the nicest thought to imagine what other countries or powerful leaders might be willing to do the day iron is no longer cheap and plentiful. Just as the Stone Age didn't end because the world ran out of stone, we can only hope that our descendants develop a new and better infrastructure that is ready before steel becomes a luxury commodity once again.

4 | Copper, Aluminum, and Titanium: From Light Bulbs to Cyborgs

WHEN MY FUTURE HUSBAND AND I went to Australia to study, we got our own car for the first time. The car was four years older than I was, and we inherited it from my then-boyfriend's Australian uncle, who, upon handing over the car, also gave us detailed instructions on how to take care of the spark plugs. When we revealed ourselves as car amateurs, he assured us that he could always help take the engine apart (and put it back together) should it start making strange noises. The car sometimes seemed tired of slogging around in the Australian wilderness, but a visit to the handy uncle always got it ready for new adventures.

The car we have today has only luggage space beneath the hood. The electric motor is hidden somewhere beneath the seats, and the dashboard is a single large computer screen. When the car makes strange noises or doesn't behave as it should, we have to call someone who can connect to the car from somewhere in the world and perhaps fix the problem by sending a software update. If not, we have to take it in to the repair shop.

As we're driving, we can monitor the car's position on a map on the big screen. I start the trip by typing in where we're going, and the car then calculates which road is the fastest and where we should stop to recharge. The rearview camera and sensors on the sides make it easy to park, but our car lacks the latest technology and can't parallel park entirely on its own. In the latest cars, you can even drive where you want without ever touching the wheel a single time, but current legislation allows this in only a few places.

In the summer of 2017, we began having the first discussions about when people won't actually be allowed to drive cars themselves. Computer-controlled cars don't sleep, argue with their kids in the back seat, or drink alcohol. As we learn to rely more on technology, we'll likely start considering it to be irresponsible to hand control over to a reckless human being. If no one is driving their own car in fifteen years, maybe my kids will never even learn how to drive, just as I myself never learned how to fix a car. Technological developments are constantly shifting the boundaries between man and machine with the aid of the elements copper, aluminum, and titanium.

COPPER IN CARS, BODIES, AND WATER

The electric car is one of the latest additions to all of the gadgets that fill up our lives. Ever since electric lighting became widespread in the 1880s, access to cheap and reliable electrical energy has had a tremendous impact on society and on everyday life for most people. Without electricity, it gets dark when the sun goes down. Then you'd have to read and write in the light of oil lamps and make food over fireplaces where you'd breathe in harmful smoke each day. In most of the Western world, these kinds of conditions belong to a forgotten time, but it's only been a few decades since my grandfather worked building power stations and laying copper wires that could conduct electricity through the northern part of Norway.

As electricity dominates more and more of our infrastructure, the amount of copper we surround ourselves with increases as a result. Copper is the most important metal for our use of electricity, primarily due to its outstanding conductivity, but also because it corrodes slowly and is quite cheap to produce. Just after World War II, the average family car contained about 150 feet (45 m) of copper wire. Today, this has increased to over 1 mile (1.5 km) in regular gasoline cars, and significantly more in hybrid and electric cars. Most of the copper is part of various electronic components that didn't exist in postwar cars.

Copper is also used for much more than just power cables, computers, and electric cars. For example, many of our water pipes are also made of copper. These can leak tiny amounts of metal into our drinking water, which in small doses isn't a problem since we need copper in our bodies regardless. Some of the most important proteins in the cell's machinery contain copper atoms, and there's enough copper in the human body to make a lump the size of a small grain of sand.

However, copper becomes toxic to the body in excessive amounts, and if you drink water that's been sitting in the pipes for a while, the copper content might make you sick. The same thing can happen if you're cooking mulled wine in a pot that you otherwise only use to boil water. Over time, the copper in tap water will form a coating on the inside of the pot, and if this coating is dissolved in the acidic wine, the festive holiday drink could end up giving you a dose of copper that's a bit on the large side.

Copper has been a part of our society since long before the Iron Age. Along with gold, copper is one of the few metals that can be found in pure metal form in nature, and was used as early as eight thousand years ago. However, copper metal deposits are quite rare, so the use of copper wasn't widespread until methods for extracting it from minerals were developed.

Copper is a soft metal and thus made for weaker tools than those later made of iron. When copper is hammered, however, it can become relatively hard. Hammering creates disarray in the crystal structures and makes it difficult for the atoms to slip past each other. When the metal is heated up again, the atoms will position themselves nicely next to each other so that the metal becomes softer and more flexible. In this way, new tools can be made from the same metal. Gradually, it became common to mix copper and tin to make bronze, and to use alloys of copper and arsenic or lead that were better suited for weapons and tools.

THE COPPER MINES THAT CLEARED THE FORESTS

Copper is a rare element in Earth's crust. Nonetheless, there are extractable copper deposits in most countries, since copper in the crust can easily be transported and concentrated through various geological processes. It's an advantage that copper thrives with sulfur, so in most cases, copper appears in sulfur-containing minerals that are easy to sort out from the ore. Thus it's possible to make money by extracting copper from deposits that are really only a few thousandths copper. While iron ore can contain well over 50 percent iron, 0.6 percent is a typical concentration in today's copper mines. This means that, for every metric ton of stone that is excavated, we're left with 13 pounds (6 kg) or so of copper and a little less than 2,200 pounds (1,000 kg) of tailings.

As with iron, large amounts of carbon and energy are used in the production of copper metal from copper ore. Timber consumption in the earliest copper production led to widespread deforestation in parts of Spain, Cyprus, Syria, Iran, and Afghanistan. In recent times, something similar has happened in my own country, on

Rørosvidda in central Norway, where copper was mined from the mid-1600s all the way up until 1977. The large forests that once grew in the area were cut down to provide fuel both for fire setting in the copper mines and for melting the copper ore. It could take up to 17 cubic feet (0.5 m³) of timber to extract copper from 1 cubic foot (0.1 m³) of rock.

As if the clear-cutting weren't enough, much of the vegetation in the area around Røros was damaged by pollution from copper production. Up until the middle of the nineteenth century, an important part of copper ore processing was done outdoors. To separate copper from sulfur, crushed ore was heaped on top of a dry wood substrate that was then lit and left to burn for a few months. The sulfur in the ore reacted with oxygen from the air and rose to the sky as gas. Up in the air, the gas reacted with water vapor and turned into sulfuric acid, which later fell back to the ground like an extreme version of acid rain. In today's copper production, methods have now been developed for stopping most of the pollution before it's released into the environment.

Copper mines and copper metal production can leave extensive traces on the landscape, but if we want to continue using electricity the way we do today, we're dependent on maintaining copper supplies for the world market. However, according to some studies, we have only a few decades left before copper production will start to decrease. At the same time, other scientists point out that the copper that is mined today is in the very top 0.5 mile (1 km) of Earth's crust, most often at the very surface. There are likely large, unknown copper deposits about 2 miles (3 km) down. If methods are developed to find these deposits along with robots that can work in deep, hot, and dangerous mines, the extractable resources could be ten times greater than those we account for today. This may even allow us to maintain our copper consumption for several hundred more years.

ALUMINUM: RED CLOUDS AND WHITE PINES

Copper isn't the only metal we use for conducting electricity. In many cases, aluminum is a good alternative; it's light, which makes it well suited for power lines. Its low weight is also the reason why the majority of my electric car is made of aluminum in an alloy with other elements that strengthen the metal without making it too heavy.

I have a close relationship with aluminum. Even though the element doesn't have any useful function in the body, human bodies still contain about as much aluminum as copper (preferably not any more, since too much aluminum can be harmful). I also hold aluminum in my hand every single day, several times a day—since the outer shell of my phone is made of aluminum. When oxygen reacts with aluminum, a layer of aluminum oxide is formed that clings to the metal beneath it like a tight, protective film. This is how the rest of the metal avoids coming in contact with oxygen, which is why aluminum doesn't corrode and disintegrate as iron does. At the factory that makes the phone shell, the oxygen levels and temperatures are controlled so that this oxidized layer becomes just thick (about five thousandths of a millimeter) and solid enough to be able to stand my handling of it.

Since 8 percent of Earth's crust is aluminum, we're talking about a pretty common element here. After iron, it's the most widely produced metal in the world (production is around 50 million metric tons each year, versus 1.64 billion metric tons of iron). Almost all aluminum is made from bauxite, a type of rock that forms in tropical areas when surface water that trickles through weathering bedrock picks up other elements and allows aluminum, silicon, iron, and titanium to remain. Most of the bauxite deposits being mined today are in Australia, China, Brazil, and Guinea.

Since it lies so close to the surface, bauxite is mined in quarries. Top layers of earth and stone are moved to the side, and the bauxite is dug out and crushed before being treated with sodium hydroxide in an enormous pressure cooker to separate the aluminum oxide from the other minerals in the ore. The residual waste is a thin red mud pumped out into enormous ponds where the sludge is allowed to slowly dry. Lye makes the red sludge corrosive, which can cause major and immediate damage to the environment if there are leaks or dam failures. The biggest event of this kind was the dam failure in Ajka, Hungary, in 2009, when ten people died—most likely by drowning—when mud flooded over the nearest village. The sludge continued into the local river, where all life was killed, and then farther out into the Danube. Fortunately, the long-term effects of the accident seem to have been minor.

In Malaysia, the authorities imposed a temporary ban on bauxite extraction in 2016 because an explosive, uncontrolled development of mining activities led to major environmental damage, including in the form of red dust clouds that blew across the landscape from dried storage ponds. The ban in Malaysia caused global aluminum production to fall by 10 percent that year and is a good example of how important it is that strict environmental regulations for the mining industry are both implemented and enforced.

For a long time, aluminum was an expensive and exotic metal, much on the same level as gold. Pure aluminum oxide has to be heated up to over 3,600°F (more than 2,000°C) before it melts. Such high temperatures not only require a great deal of energy but also make it almost impossible to find materials with which to build a furnace. At the end of the 1800s, metallurgists discovered that aluminum oxide's melting point can be lowered to around 1,800°F (roughly 1,000°C) by combining it with cryolite, a fluoride mineral. Without this breakthrough, we wouldn't have aluminum in cars, phones, or beer cans today.

To get this molten mixture of aluminum oxide and cryolite to react with carbon to transform aluminum into metal, the melt must be plugged into an electrical circuit in which the electrons are forced to move from carbon to aluminum. This process requires large amounts of electrical energy, which is why aluminum oxide from bauxite mines in the tropics is sent to places where cheap electricity is readily available.

As a child, my family often visited the beautiful Utladalen valley, which lies in the western part of Jotunheimen National Park in southern Norway. The end of the trip went through an old forest up to a mountain plateau called Vettismorki. Here, the largest pine trees are completely white. My parents told me that these trees had died due to emissions of fluorine gas from the aluminum plant in Årdal. I'd always thought this was a strange story—that fluoride, which I took for my teeth in the form of small tablets with smiley faces each night, could kill such large trees, and that fluoride had anything to do with metal at all.

The aluminum plant in Årdal municipality at the innermost part of the Sognefjord was started up by German occupiers during World War II. After the war, the plant was taken over by the Norwegian state and is now operated by Norsk Hydro as one of the world's most advanced aluminum plants. The availability of cheap hydropower makes Norway an attractive place for producing aluminum, and today Norway is the world's eighth-largest aluminum producer.

The start of aluminum production in Årdal in 1949 had an immediate effect on the livestock in the area. They suffered major damage to their teeth and skeletons and became so weak they had to be driven up to the mountain pastures. This rather obvious link between industrial emissions and harm to nature and livestock led to several lawsuits during the 1950s in which the plant ended up having to pay damages to local farmers. All the attention this brought contributed to the beginning of Norwegian environmental policy with the founding of Røykskaderådet (the Smoke Damage Council—which later became the Norwegian Pollution Control Authority, and later the Climate and Pollution Directorate) in 1961.

As I learned from my parents, it was fluorinated gas from the molten cryolite that damaged coniferous trees and animal teeth and skeletons. We give our children fluoride mouthwash and use fluoride toothpaste because small amounts of fluoride can enter the surface of the crystals that our teeth are made of and make them stronger. On the other hand, if fluoride levels become too high, we can't form the right type of crystals and our teeth will be damaged.

It wasn't until the 1980s—around forty years after the plant opened—that cleaning systems were installed that were good enough to prevent damage to the coniferous forest. Today, the purification systems capture most of the fluorine and send it back through the extraction process. Fluorine emissions from Årdal are still affecting the teeth of local deer, but the environmental impact of the plant is almost negligible in comparison to how conditions were a few decades ago.

USING WHAT WE'VE ALREADY USED

At today's rate of extraction, there's enough bauxite to supply the world with aluminum for about three hundred more years, but when bauxite deposits start to run out, we'll need to begin extracting aluminum from other minerals. Aluminum is so common in Earth's crust that we should be able to extract it as long as we have enough energy to spare.

Aluminum can also be recycled. Producing aluminum from recycled materials requires only a fraction of the energy that goes into producing new metal. This means that aluminum is one of the most recycled materials today; worldwide, over 60 percent of all discarded aluminum is recycled. Still, less than half of the aluminum on the market comes from recycled material, but in just a few decades, recycling is likely to be more important than mining.

Metals are generally well suited for recycling. When melted, they behave just like new material. To be fair, it can be quite difficult to split alloys back into their individual components, which is why metal recycling must be sorted carefully to ensure that the different alloys aren't mixed together. This would ultimately end in undesirable properties in the finished product. There are good chemical methods for analyzing the content of alloys, but the sorting job is both simpler and cheaper if the various components are well labeled and easy to disassemble.

My phone is made of much more than just aluminum. An average mobile phone contains more than thirty different elements— one-third of the eighty-three nonradioactive elements found here on Earth. The electronics in phones are built from crystals of pure silicon, which are combined with tiny amounts of elements such as phosphorus, arsenic, boron, indium, and gallium to create the electrical components used to control signals and store information. The electrical connections are made of silver, which conducts electricity best; gold, which never rusts; and copper, which is the cheapest. The glass on the screen is made of silicon and oxygen combined with aluminum and potassium. When I touch the screen, electrical signals are transmitted into the computer inside the phone because the glass is covered with a layer containing indium and tin so thin that I can see right through it.

If we want to be able to continue creating ever more sophisticated computers and communication tools in the future, researchers will need to develop new and better methods of separating all of these elements from each other after use so they can be reused. We've spent thousands of years perfecting methods for making metals from stone. Now, all the knowledge we've acquired needs to be used to figure out how to extract the metals we need from the scrap heaps of civilization. Discarded cars and phones could be the gold mines of the future.

THE TITANIUM IN A MOUNTAIN

My car's undercarriage is made of titanium, a light metal that's much stronger than aluminum but also much more expensive. It's therefore only used where a combination of low weight and high strength is extra important. Titanium is important not only in light vehicles that use little energy but also in spacecraft such as satellites, which we depend upon so that our cars and phones know where they are, and to keep track of the weather, ice cover, and vegetation down on Earth.

As a metal, it may not be in outer space that titanium really comes into its own but rather inside the human body. Sometimes we also need spare parts. As early as ancient Roman times, people had artificial teeth made of cast iron, and the first successful artificial hip was inserted in 1938. Implants give a lot of people an increased quality of life for many years.

It's important that implants are made from materials that can function inside the body over a long period of time. They can't rust or corrode and break into pieces, and they can't release any substances that are harmful to the body. Gold, silver, and platinum meet these criteria, but these metals bend easily under applied stress and are therefore poor substitutes for teeth and bones. Stronger metals such as iron, brass, and copper can corrode and irritate the body, although ancient Roman cast-iron teeth appear to have worked amazingly well. Among all the metals, titanium alloys function the best. Titanium is strong and lightweight and can remain in the body for a long time without weakening or causing any unwanted side effects. This is one use of titanium we'll certainly want to continue taking advantage of for a long time.

However, the majority of titanium that's extracted isn't used in its metal form. Almost all of it—about 90 percent—is used as titanium dioxide because of its extra white color. Titanium dioxide has replaced lead in many kinds of white paint, which makes the use of titanium good news for the environment. The problem is that paint is one of the most difficult things to recycle. Just as we've lost much of the gold used for gilding surfaces over the years, it's in paint's nature to wear off—so we have to repaint. Worn paint turns into dust that is carried by wind and weather into the seas. The titanium we use for paint today won't be available to be used for space shuttles or implants in the future.

Norway has been extracting titanium for over a hundred years. In the rest of the world, titanium is often extracted from sand, where light minerals are washed away while the heavy titanium minerals remain (because even though titanium is a light metal, the minerals it's included in are heavier than most other minerals in the sand). In Norway, however, we have some of the world's largest deposits of titanium in solid rock. To extract it, the rock must be crushed into tiny particles—less than half a millimeter wide—and mixed with water. Then, most of the titaniferous minerals can be sorted out using magnets and gravity. The very smallest of grains are captured by mixing soap with the sludge and making it foam, such that the titanium minerals attach to the bubbles and can be scraped off the top.

After processing, the titanium ore is converted into marketable titanium oxide, but large quantities of sludge are also produced that need to be put somewhere. When the Tellnes mine in western Norway opened in the 1960s, the sludge was deposited on the bottom of a nearby fjord called Jøssingfjord. First, the shallow areas of the fjord were filled. Subsequently, the mine owners wanted to start depositing sludge in the 300-foot (90 m) deep trench called Dyngadjupet farther out in the same fjord. This was met by massive protests from both environmental protection organizations and fishermen, who even occupied the office of the Minister of the Environment in 1987. Despite the protests, the permit was granted and mine sludge was pumped into Dyngadjupet for ten years.

In 1994, Tellnes switched to disposal sites on land, where they pump two million tons of sludge into a dammed valley each year. When the surface dries out in the sun, the wind can swirl up large clouds of dust, and when it rains, rainwater trickles down through the sludge. Chemical reactions between rainwater and minerals release heavy metals such as nickel, copper, zinc, and cobalt. The water flowing out of the landfill carries this pollution into streams and farther out to the fjord—and this process will continue more or less for all eternity.

Sludge deposited into the sea doesn't emit heavy metals in the same way, both because the chemical composition of seawater makes the minerals more stable and because the seawater in the deposit site doesn't move as much. The problems with sea deposits are related more to the particles themselves. An obvious problem is that the seabed gets covered up, wiping out all life there. In addition, the smallest particles pumped into the fjord immediately sink to the bottom. If they join the current out of the fjord, this can create problems for ocean life in a larger area. Small particles can settle in fish gills, and the sediment makes the floodwaters darker, thus altering the entire food chain.

When mining operations end and the sludge is left to rest, life on the fjord floor should theoretically return. After thirty years, however, Jøssingfjord still bears distinct traces of the deposits. Ultimately, society has to weigh several interests against each other: Do land deposits cause more environmental damage than sea deposits? Will the financial gain from titanium mining be enough to accept the consequences all of the extraction will have on the environment? And we also need to ask: Should we sell the titanium as paint today, or should we use it for implants in the future?

THE CYBORGS ARE COMING!

Just a few years ago, I was looking things up in the yellow pages and a physical dictionary, reading paper maps, making plans in advance with people about when we should meet, buying tickets in kiosks or automated machines, checking booklets with bus and tram timetables, carrying around a camera to take pictures, doing calculations on a pocket calculator, going to the bank, measuring time with a stopwatch, using an alarm clock, and writing appointments and plans in a notebook. I now do all of these things on the little computer I carry in my pocket all the time: my mobile phone. Like many others, I often feel a kind of impulsive urge to hold my phone in my hand and do something with it, such as checking my email or going on Facebook, even though I know I could be doing something more useful with my time.

Now that we've become so attached to a computer that we want it close to our bodies at all times, we start to wonder if it's even really necessary to have it separate from the body at all. There are already options you can attach to your body, for example, a "smart watch" worn on your wrist. You can also get glasses that provide you with information right in your field of vision so you don't have to bend over a screen. These glasses may also even have a built-in camera that you can use to take pictures of what you're looking at whenever you want to. My cat has a chip under his skin that lets him open the cat door, and in some American workplaces, employees can implant a similar chip in their hand that they can use to clock in and out of work or pay for lunch in the cafeteria. This chip is a tiny little computer with information that can be changed through electrical signals from outside the skin.

The human body also uses a form of electricity. We've known since the eighteenth century that electrical signals in our nerve cells are used to control muscle movements. By measuring and controlling these signals, we can therefore theoretically both investigate and control what is happening in the body.

The pacemaker was the first implant used to measure and send electrical signals in the body's systems, in this case the muscle cells of the heart. When a pacemaker detects that the heart isn't beating as it should, it sends a signal that forces the heart to beat with a steady rhythm. Swedish engineer Arne Larsson was the first patient to receive a pacemaker, in 1958. Even though he had to have a replacement put in after only eight hours and went through twenty-five operations to replace or repair pacemakers before he died in 2001, these have quickly developed into machines we can truly trust. Today, we also have retinal implants that help the blind to see, cochlear implants that let deaf people hear, and electrodes that can be implanted deep inside the brain to treat diseases such as Parkinson's, chronic pain, epilepsy, anxiety, and depression. These electrodes send electrical impulses into the brain's own signaling system and can therefore control some of what the brain does itself. Such electrical connections located inside the brain can be extremely precise, but in many cases, electrodes sitting just inside the skull or even on the surface of the head can suffice.

Electrical circuits can also be connected to nerve cells or muscle cells that have contact with the body's central nervous system. In this way, the body's own signaling system can be used to control machines outside the body, such as a hand prosthesis. The brain has an amazing ability to learn how to operate such external machines. It doesn't use the same nerve paths as it would to control, say, a real hand. It's enough to simply look at the artificial hand or to understand how it moves for the brain to build connections between nerve cells that enable it to control the machine as if it were a real body part.

Direct links between machines and the body's signaling systems can also be used in the opposite direction: Signals from the outside can affect the brain or muscles. Insects have simpler signaling systems than we do, and systems have been built that can allow you to remotely control beetles, grasshoppers, and moths with implanted electrodes that are connected to a small computer inside your head. You could build an army of remote-controlled grasshoppers if you had the need for small machines that could take pictures or get into confined spaces. Animals with more complex brains such as rats and pigeons can be controlled by stimulating the brain's punishment and reward systems. In this case, electrodes are connected directly to nerve cells, or signals are used to release chemical substances that are absorbed by brain cells.

Cyborgs, a kind of hybrid between humans and machines that often have extraordinary abilities, are something we know from films and literature. In a way, however, we could say that people with pacemakers and retinal implants are actually already cyborgs themselves, and we have every possibility of taking this development even further.

Today, it's almost inconceivable not to let children have their own phones at some point before they start middle school. When our children grow up, it might be unthinkable for them to be drivers in their own cars. When my children have children themselves, it might be routine to get machine implants that provide them with various benefits: health monitoring, better sight and hearing, or even the ability to communicate with the outside world, pay bills, and send messages without having to use an external machine.

THE FUTURE OF MACHINE PEOPLE

Electronic components are getting smaller and smaller. The computer in my phone is more powerful than the one my father used at work when I was younger—and that was as big as a refrigerator. Today, we've learned about how materials function all the way down to the atomic level, letting us build machines that are so small we have to use advanced microscopes to see them. We can even send computers and robots into our veins and cells if we want to.

You might think that the development toward smaller and smaller machines would be good news for those who are worried about running out of materials in the future; after all, small machines require fewer raw materials. This is one of the arguments that can be used to say that we can continue growing and developing as a civilization without increasing the burden on the world's resources. Small devices also need less energy to function. In the future, small machines in our bodies could even harvest the energy found naturally in our bodies so that they can work without batteries that need to be recharged.

However, making small things comes at a price. The smaller something is, the cleaner it needs to be to work. While a large radio made of metal parts big enough to hold in your hand can work well with a fair amount of impurities, every single atom becomes important when the components are so small that they only consist of a few atoms themselves. In today's electronics industry, production takes place in laboratories that are so clean that a single grain of dust can cause serious problems. This can only work with advanced ventilation and filtration systems, which in turn require a lot of energy and regimented control of everything going on in the plant.

There is a big difference between getting something fairly clean, very clean, and ultra clean. Impurities from a substance can, for example, be removed by distillation, which works because all pure substances have their own boiling temperatures. When alcohol is produced by distillation, a mixture of water and alcohol is heated until the alcohol becomes gaseous, while only some of the water does. When the alcohol gas is collected and cooled down, it will condense and turn into liquor, which contains some water. To remove as much of the water as possible, this process must be repeated several times. A lot of energy goes into evaporating the alcohol in each round, and in each round, some of the alcohol will inevitably be lost. The same principle applies to all other materials that need to be purified. Even though the tiny machine you're left with in the end consumes little energy and weighs almost nothing, it is concealing a tremendous consumption of both energy and chemicals during the production phase. The energy used for chemical separation (all of the processes in which substances are separated—although to be fair, not just for making electronics) constitutes about one third of the energy used in the entire global transport sector.

However, there are other ways of making small things. Bacteria and other living organisms are, in a way, also small machines. Researchers can already change gene material in simple organisms to get them to produce certain chemical substances. Some bacteria are able to make electrical connections only a few atoms thick—entirely on their own. Work is now underway to figure out which genes control the production of these materials. In the future, we could use this knowledge to design our own electronic components by developing bacteria specifically for this purpose.

In the future of electronics, in interacting with other living organisms, we ourselves could be computers, and parts of computers could be made less of metals extracted from Earth and more of living organisms extracting their energy from sunlight, and not fossil sources. This still requires many more years of research and advanced, expensive facilities.

We'll never be able to get bacteria or plants to make all the electronics we need, and this applies not least to space travel. Materials that will be used outside Earth's atmosphere need to be able to withstand a great deal of stress. We know that we can get sunburned or even skin cancer if we spend too much time in the sun. This is because the most energy-rich part of sunlight (ultraviolet) can destroy chemical bonds in the molecules that make up our skin cells. Luckily for us, photosynthesis by plants has given us a layer of ozone in the atmosphere that prevents most of the hazardous radiation from reaching Earth's surface. Beyond this layer, however, radiation is much more powerful. We can't expect organic molecules in bacterial electronics to be able to withstand a trip to space.

In orbit around Earth, lightweight materials that can withstand radiation, cold, and heat are required. Here, light metals such as aluminum and titanium will be especially important. In addition, we'll use a lot of what we call "ceramic materials"—which we'll take a look at now.

5 | Calcium and Silicon
   in Bones and Concrete

I LIVE IN A BRICK house. The foundation is cast in concrete, the walls are insulated with fiberglass, and I look outside through a glass window. I have tiles on the wall above the kitchen counter and on the bathroom floor. In the bathroom, both the sink and toilet are made of porcelain, which is the same material used in the cups and plates in my kitchen cabinet. In my mouth, my teeth are covered with a hard layer of enamel, and beneath that layer my teeth are made up of crystals that contain calcium, phosphorus, and oxygen, as well as a bit of silicon. When these elements work together to share their electrons (the way they like to), they can form hard, but brittle, ceramic materials.

Although we might not pay much attention to them, ceramic materials play just as large a role in our daily lives as metals. In addition, ceramic materials form important components of both aerospace technology and in many of the smartest and most advanced machines we surround ourselves with. Such components must be made in clean, advanced laboratories, where their composition is controlled almost atom by atom. Some of them can also be used to produce electricity through small temperature differences. Perhaps these will be able to harvest enough energy to power the small computers we'll put in our bodies in the future? No doubt about it: Ceramic materials will play an indispensable role in the development of future technology.

HARD AND BRITTLE

The ceramic materials form a diverse group, but they have some important features in common. For one, they're all hard and can withstand a great deal of stress; it isn't without good reason that the world's largest structures are made of concrete, or that teeth are covered with enamel. At the same time, they're brittle, which means they'll break before they bend if the load becomes too much. If you bite into something that's too hard, a piece of your tooth enamel can crack. Very few ceramic materials can conduct electrical current, which is why ceramics, glass, and porcelain are used as insulators in high-voltage lines to ensure the voltage doesn't pass from one conductor to another, or into the high-voltage pylon and into the ground. Without such insulators, we wouldn't be able to move electrical power around in society as we do today. Ceramic materials are also quite poor at conducting heat, which is why I can easily hold a porcelain mug of steaming hot tea while I would have burned my hand if the cup had been made of metal.

Ceramic materials behave fundamentally differently from metals. We produce metals by taking elements that were in good partnership with others and forcing them to accept extra electrons. When the atoms then come together to form a piece of metal, they feel no further "responsibility" for the electrons they were saddled with. In a way, these electrons become like children on summer break, running freely around the material. This makes the metals conduct both electricity (electrical current is nothing but electrons moving through a material) and heat, which is also easier to transport when the tiny components of the material can move around. The structure of atoms that line up in a sea of moving electrons also makes it easy for the atoms to slip past one another, making the material bendable and stretchable.

Ceramic materials also consist of tiny crystals made up of atoms neatly arranged in rows that stick together. Ceramic crystals differ from metals in that they have distributed the electrons between them (just as they prefer) so that all of the electrons are carefully "watched" and have little freedom to move between the atoms. This causes the atoms to attach extra tightly to their neighbors, making it virtually impossible to get them to slip past one another when the material bends. That is why these materials can withstand heavy loads but break when the load becomes too great.

MOLDING WITH CLAY

The simplest form of ceramics is also the oldest. The art of working with clay has been with us since the beginning of human history and remains an important part of our culture. When I went to elementary school, ceramics was one of my favorite activities in art class. Actually, I'm not even sure if we called it "ceramics"; I think we just called it "forming things with clay." We each got a lump of moist, reddish-brown clay that smelled like soil and used fun tools to shape it into an animal or a little bowl before firing it in the oven and turning it into a gift for Mom and Dad.

Clay is a word with several meanings. In everyday speech, it's used to define heavy, dense soil. Technically, clay is classified as a soil in which all of the particles are as small as grains of dust. The crystals that build up clay soil are called clay minerals that form from solid rock when it's broken down and weathered in contact with water on the earth's surface. At the atomic level, the clay minerals consist of strong, ultrathin layers of silicon and oxygen. Between each layer, there is usually aluminum, calcium, or iron, but the clay minerals also have their own ability to absorb both water and other elements between these layers.

In moist clay, these minerals cling both to each other and to the water between them. This makes it possible to mold moist clay into all kinds of intricate shapes. When clay objects are fired in a hot oven, the water evaporates, and the clay minerals stick together so well that the finished ceramics become as hard as stone.

Between 10,000 and 14,000 years ago, ceramic production evolved from simple figures to useful objects such as bricks, tiles, and jugs. The fired clay itself isn't impermeable enough to hold water or oil over a longer period of time, however. In order for ceramics to become truly waterproof, the potter has to melt the outermost layer. Perhaps the first glazed clay jugs were created by accident once when the oven got too hot. Later, potters learned to glaze ceramics by covering the surfaces of the fired objects with a powder that lowered the melting point of the clay minerals it came into contact with, causing the clay surface to melt and become glassy upon a second firing. In ancient times, such jugs were used to store wine and oil. Today, there are ceramic cups and saucers with glazed surfaces in my kitchen cabinet. The porcelain soap dishes and tiles in my bathroom are also a kind of glazed ceramic made of a special type of clay mixed with powder from quartz and feldspar in a process that was developed by the Chinese in the 600s.

THE MESSY ATOMS IN THE WINDOWPANE

Having mastered the art of making glazed ceramics, it seems like a natural next step to melt down the entire ceramic object to get pure glass. However, this is much easier said than done. Glass production requires such extreme conditions that it took thousands of years from when humans started glazing ceramics until they could produce pure glass. The oldest human-made glass that's been found is about 4,500 years old.

Glass can be found in nature. It can be formed in volcanic eruptions when molten rock is cast into the air and cools down before the atoms are able to crystallize. When tectonic plates scrape against each other and cause earthquakes, the friction can produce so much heat that a thin layer of rock melts before solidifying quickly when the movement stops. We can see the result in the form of thin veins of glassy material in the rock. The temperature can also get high enough to melt rocks when enormous meteorites collide with the surface of our planet. What all of these processes have in common is that glass is formed when the temperature is high enough to melt all of the minerals in the stone. When the melt cools down quickly enough that the atoms aren't able to find their way back to ordered crystal structures again, they instead end up in random locations. This is what glass is: a ceramic material in which the atoms are all helter-skelter.

The glass in my windows is made of pure sand. This sand should only consist of the minerals quartz, which contains silicon and oxygen, and feldspar, which contains aluminum as well as potassium, sodium, calcium, or barium. The mixture of quartz and feldspar won't melt until the temperature reaches about 3,600°F (roughly 2,000°C)—and it's almost impossible to build a furnace that can withstand such a high temperature. Sodium carbonate—which is extracted from salt mines—is therefore added to lower the temperature to about 1,800°F (1,000°C). In addition, crushed limestone must be mixed in to prevent the finished glass from dissolving in water. When this mixture is heated, carbon dioxide is released from both the sodium carbonate and the limestone and disappears into the atmosphere, meaning that the finished glass weighs less than the sum of all the raw materials that went into making it.

It's fascinating to watch glassblowers at work. Over and over, they suspend the glass in a furnace more than 1,800°F (1,000°C) to keep the glass smooth and malleable. Most of the glass we use in daily life is made by machines, however. The drinking glasses in my kitchen cabinet were made in molds. The large glass panels in my living room windows were made by pouring molten glass over a tub of molten tin. Here, the glass flows out in a uniform layer, flatter than it could have done in any solid mold. The glass in my car windshield was cooled extra quickly to make the atoms in the finished material stand in tension against each other. This makes them more difficult to tear apart so the window doesn't shatter simply from a flying pebble.

The glass's color and other properties can be adjusted by adding tiny amounts of different elements. The green color in many beer bottles comes from iron oxides. When I make lasagna, I use an ovenproof glass dish with a dash of boron oxide, which allows the glass to withstand temperature changes in the oven without cracking. By adding lead, you get easy-to-cut glass that gives a clear tone when you tap a spoon against it. However, you should always be careful about drinking from the very finest crystal glass; you're better off not having too much lead in your system.

Glass isn't just found in windows, windshields, and drinking glasses; it's also used in some of our most advanced communication systems. By adding the right elements, manufacturers can control how light travels through the glass, and glass can be stretched into thin fibers and used to transmit light over long distances. Cables with long bundles of such optical fibers have been buried along our roads on a large scale in recent years to connect our homes to each other and the rest of the world through the internet. In the future, small components of specialized glass will be used to transmit information in the form of light in many of the instances where electronics and metals are used today.

As with metal alloys, it's hard to separate out the individual elements that are mixed in the glass. It is therefore important that different kinds of glass are sorted before being melted for reuse. Even a small amount of the wrong kind of glass in a furnace can be enough for all contents to have to be discarded. Apart from this, glass is well suited for recycling. It can be melted down and reused again as if it were new.

FROM ALGAE TO CONCRETE

The very first ceramic material that was used—and perhaps even the very first tool *anyone* used—was a rock someone picked up from the ground. We live on a planet of rock, and the rock we walk on, climb on, and blast tunnels through is also a ceramic material. Rocks come in different variations, and some of them are better suited for making tools and weapons than others. Flint is an example of a stone that's easy to make sharp tools with, and it can be picked up from beaches or excavated from limestone in parts of Denmark and Sweden. The US has a deposit rich in flint in the Appalachian foothills of eastern Ohio. Norway only has flint that was transported there via ice in colder times, which made it a sought-after commodity for the early people of my country.

Stone has also been used as a building material since the beginning of time. With the right technique, you can create fairly advanced structures by laying stone upon stone. Nevertheless, the range of choice of building types becomes a bit limited when gravity is the only thing holding the structure in place. When people discovered how large and small stones can be bonded together with mortar or cement, it opened up opportunities for far more varied structures.

The first of these binding agents was produced from limestone, a rock we find in many places here on Earth. The soft, porous chalk stone in the White Cliffs of Dover, the Italian marble covering the opera house in Oslo, and the stone that makes up the Empire State Building are all types of limestone. Limestone is made up of the remains of biological ceramic materials. In the ocean, there are tiny algae floating around on the surface of the water. Some of these make shells—a miniature armor of crystals made of calcium and carbon dioxide that has dissolved in the seawater. Algae that lived millions of years ago did the same, and when these algae died, they sank to the bottom, where the shells lay like a chalk-white dust on the bottom in ever-thicker layers along with remnants of larger shells and corals. Later, the climates and oceans changed. The shells on the seabed were covered with sand and mud, and over the next few millions of years, they were gradually pushed down into Earth's crust and converted into solid limestone.

Since algae and other marine animals built their shells from carbon dioxide dissolved in water, which had in its turn come into the seawater from the air, limestone represents a huge stockpile of carbon dioxide from the atmosphere of the past. If limestone is heated to over 1,500°F (roughly 800°C), the crystals will break down and the carbon dioxide will disappear back into the atmosphere. Left behind is a powder of calcium and oxygen, which reacts powerfully and produces a great deal of heat if it comes in contact with water. What's formed in this process is called slaked lime (calcium hydroxide).

Slaked lime absorbs the carbon dioxide from the air it comes in contact with and hardens into a material that is really nothing more than a new limestone. Mixing slaked lime with fine-grained sand gives you the simplest form of lime mortar, which you can use to bind stones into formations, or plaster, which can be used to cover stone walls. Mixed with sand and gravel, slaked lime was the very first form of concrete.

No one knows how humans discovered they could use heat and water to transform limestone into a new, human-made stone. An ordinary campfire isn't hot enough to break down limestone. Perhaps it was a lightning strike or a forest fire that transformed the limestone on the ground into the reactive slaked lime powder, which then aroused the curiosity of the people who discovered how the powder hardened in contact with water. The oldest discovered remains of concrete floors made with slaked lime are as much as 12,000 years old. This means that they were made before humans became settled and started farming—at around the same time that ceramic technology started gaining momentum.

We might even be able to say that industrial history started with calcination, the very first chemical industrial process. This was the first time humans were producing and managing large quantities of hazardous chemicals. The fine powder of calcium oxide is quite reactive and could have damaged the skin and eyes of those who had to work with it. It must have required very good planning and cooperation to carry out these advanced operations.

VOLCANIC ASH IN THE COLOSSEUM

About four thousand years ago, the Minoans, a seafaring and trading people in Crete, had one of the most advanced civilizations in the Western world. It lasted until about 1640 BCE, when the Mediterranean island of Santorini exploded in an enormous volcanic eruption, triggering a tsunami that swept over and almost wiped out the Minoan trading towns. After the disaster, the Greeks came in and took over the old Minoan territories, and both the Minoan written language and culture sank into oblivion.

However, our interest in the Minoans here isn't due to their legends or mystical writings, but rather their unique concrete technology. The Minoans were likely the first people to develop what is called hydraulic cement. While masonry of slaked lime and water doesn't harden until the cement is allowed to react with the carbon dioxide in the air, hydraulic cement becomes hard as rock by reacting with water. This makes the hardening process faster and more controllable, and it also means that concrete can be used for underwater structures and allows for much stronger constructions than the ones made of pure limestone.

The Minoans made their concrete by mixing limestone with volcanic ash, which is found in abundance in the Mediterranean. Volcanic ash consists of tiny particles of silicon and oxygen, and these particles react easily with other substances both because they are small and because they were formed by glowing hot material that cooled down very quickly after the volcanic eruption. When the mixture of ash and slaked lime is mixed with water, the calcium, silicon, and water will bind together into a strong material that fills the voids between the grains of sand in the concrete. In this way, concrete becomes both strong and waterproof.

When the Minoans disappeared, so did their knowledge of hydraulic concrete. It would take more than a thousand years for the technology to reappear, when the Roman inhabitants of the area surrounding Naples rediscovered the art around 300 BCE.

The Romans became masters of using concrete. Among the many striking examples of Roman concrete we can still see today are the Colosseum, the Pantheon, and numerous remains of aqueducts and roads. It's incredible that this concrete has held up even after two thousand years.

The first large-scale project in which concrete played the main role was the construction of the port of Caesarea, in the northern part of what is present-day Israel. Herod the Great, who was king of Judea until 4 BCE, wanted to build a port large enough to accommodate the Roman grain ships. The problem was that the coastline he had at his disposal was hardly more than a single long beach, without any natural archipelago or estuary that could provide shelter in bad weather. Since he also lacked a local source of good building stone, it would have been virtually impossible to construct a proper port without hydraulic concrete.

The construction of the harbor required large quantities of timber, both for the construction of the port itself and as fuel for the kilns and furnaces that produced the clay pots to transport the slaked lime. Most of the timber was sourced from newly conquered areas north and south of the Danube, including the kingdom of Dacia (today's Transylvania), where the Romans also exploited the rich gold deposits. Herod must have had hundreds of lime kilns running day and night for several years in order to produce enough concrete for the construction project. Volcanic ash was sent on large ships from Naples. The slaked lime was mixed with water, poured into clay pots, and loaded onto ships. When the port was completed, it made Caesarea the largest and most flourishing city in Judea—the same size as Athens at that time.

Today, the harbor lies at a depth of 40 feet (12 m). The constant earthquakes along the coast of Israel have caused it to sink gradually into the sea over the centuries.

CONCRETE THAT SCRAPES THE CLOUDS

With the fall of the Roman Empire, knowledge about concrete production was lost once more, and it wasn't until the 1700s that the use of hydraulic cement started to pick up again. In the northern parts of Europe, there weren't any large deposits of volcanic ash like you can find in the Mediterranean, but masons discovered they could make cement that hardened in water if they fired limestone together with clay. The clay is heated to such a high temperature that the minerals disintegrate and a material that can easily react with water is created. The remains of the burnt clay have about the same qualities as volcanic ash, which is also a mix of siliceous minerals that have been heated up vigorously.

Almost all of the cement made today is based on variations of a recipe patented by British mason Joseph Aspdin in 1824 under the name "Portland cement." The name refers to a popular, grey-white building stone: limestone from the island of Portland outside of Dorset, England. The new cement was marketed as "just as hard as Portland-rock." Portland cement was made by heating a mixture of limestone and clay to over 2,650°F (1,450°C). At this temperature, the minerals in the raw materials are broken down and create a rather unstable material that reacts quickly when it comes in contact with water. The last step involves mixing in additives such as gypsum powder and ash from various industrial processes in order to control how much time the cement needs to harden, how thick or liquid it will be before it hardens, and how strong the finished product should be. To make concrete, cement is mixed with sand and gravel before casting.

The reactions between cement powder and water make the water inside the cement quite basic (or alkaline, the opposite of acidic). This makes Portland cement an excellent material to use together with steel. The basic environment inside the concrete, which is maintained until all of the cement has reacted, creates a dense film around the steel that prevents it from rusting in contact with water. Today, the vast majority of the world's structures are constructed of steel-reinforced concrete. The combination of steel and concrete provides such strength and flexibility that we can create virtually any structure we could possibly imagine, including skyscrapers towering over 3,000 feet (900 m) in the air and dams larger than anything the world has ever seen.

At the same time, steel reinforcement gives our concrete structures a much shorter life span than those the ancient Romans built. When all the particles in the cement have been able to react with water, the water becomes less basic, and the protective coating on the steel weakens. To be fair, this may take a very long time. In large constructions like the Hoover Dam (which was completed in 1935), these reactions are still going on to this day, meaning that the steel is still protected. In constructions of a more normal size, it's very important that water doesn't come into contact with the steel reinforcement.

Newly cast concrete is waterproof and protects its steel well. However, in the months and years after production, chemical reactions take place within the concrete that cause it to crack. This makes the concrete weaker and more porous, and allows water to penetrate it. When water reaches the reinforcing bars, they start to rust. The rust takes up space and pushes on the concrete around it, allowing even more water to come in. If these processes are able to continue undisturbed, what was once a strong bridge can be transformed into a crumbling structure that could collapse without warning. It takes time for damage to the steel reinforcement to become visible from the outside, but when you see a rusty spot or large, open cracks, the deterioration has already become extensive.

Today, concrete structures have a projected lifetime of about a hundred years. Experience has shown that damage can arise even earlier, especially in areas where the structures are submerged in seawater or are sprayed by the sea, since the salt in seawater accelerates corrosion. Our infrastructure—roads, bridges, dams, airports, building foundations, drainage pipes, and cisterns—consists of an ever-increasing number of structures that will require maintenance or replacement in just a few decades. At the same time, huge quantities of concrete are being produced that can't be recycled, since it's not possible to make new cement from used cement.

The problems with rust in reinforcement means that the concrete industry is constantly on the lookout for alternative reinforcing materials. Fibers made of carbon, glass, or plastic could be well suited for some applications, but there are still no alternatives that can compete with steel when it comes to both price and strength. Surface-treating concrete can make it harder for water to penetrate the surface, thus extending the life span of structures. There have also been experiments with self-repairing concrete, where water that enters through cracks initiates reactions that seal the cracks up again.

IS THERE ENOUGH SAND?

We're producing new concrete all the time, all over the world. Until very recently, sand and gravel for concrete production were collected from places close to the construction site, either from riverbanks or from sand and gravel pits that had been established where rivers or glaciers had deposited loose materials for thousands of years. River gravel is the best option for use in concrete. Sand and gravel from salt water must be rinsed well, since residual salt will cause the steel reinforcement to rust. Desert sand, of which there are huge amounts in the world's dry regions, isn't very well suited because it is too round, too polished, and too well sorted. To make strong concrete, it's important to have both large, strong grains as well as small grains that can fill the gaps between the large ones. Round, polished grains make it harder for the cement to properly adhere to the surfaces of the grains.

Over the past twenty years, concrete production has quadrupled in China, while it's risen by over 50 percent in the rest of the world. This has eaten into the natural deposits of sand and gravel. In Europe and the densely populated areas in the eastern US, most of the sand and gravel for concrete production is now made by crushing solid rock. Other places in the world are getting more and more sand from beaches and the bottom of the ocean.

The extraction of sand and gravel isn't without consequences for the environment, no matter where it's been sourced. When loose materials are picked up from the riverbed, it changes the river's flow pattern. This can cause more erosion both upstream and downstream of the extraction site and lead to the river changing course. Removing a lot of sand from the bottom of a river or lake can result in lower water and ultimately groundwater levels in the surrounding area, making agricultural areas more prone to drought. Extraction of sand from the seabed destroys life there, and small particles that are swirled up and released from the extraction boat can cloud the water and disrupt life there as well.

Sand and gravel make up between 70 and 90 percent of the 50 to 60 billion metric tons of solid material that we humans extract from Earth's crust each year. Of this total, 180 million metric tons go to industry, including the production of glass, ceramics, and electronics, and the rest is used in construction. We humans extract more than twice as much sand and gravel each year as what's moved by all the rivers in the world. In this way, we're actually shaping Earth's surface more than all of the natural geological processes do.

The city of Dubai in the United Arab Emirates has been responsible for some spectacular construction projects in recent years. The building of the artificial "Palm Islands" and "World Islands" off the coast necessitated 635 million metric tons of sand. Now, all the good sand deposits in Dubai have been used up. The concrete in the 2,700-foot-high (828 m) Burj Khalifa Tower, completed in 2010 and still the world's tallest building a decade later, was constructed with sand imported from Australia.

Singapore is a small island with many inhabitants. The population tripled between 1960 and 2010, and in order to make room for more infrastructure, the country has expanded into the sea. Singapore is now the world's largest sand importer and the country that uses the most sand per capita. The sand is imported from neighboring countries Indonesia, Malaysia, Thailand, and Cambodia, leading to conflicts over illegal extraction and sales as well as the alleged disappearance of twenty-four sand islands in Indonesian waters.

It seems that something as apparently trivial as sand and gravel can actually have geopolitical consequences! Good sand resources aren't coming back, so we'll need to expect more conflicts of this nature moving forward if we're going to continue building with concrete, and we really don't have any suitable alternative materials to build and maintain our infrastructure. We already produce twice as much concrete as we do of all other building materials combined. There isn't enough of anything to replace concrete.

LIVING CERAMICS FACTORIES

Among all living beings on this planet, it's only we humans who use industrial processes with extremely high temperatures to produce and shape our materials. There isn't any other creature that uses metal tools. Still, animals, plants, insects, and bacteria make some of the strongest materials we know of.

Living organisms have the ability to produce advanced ceramic materials. For example, the cells in our body make bones and teeth. The spikes of the sea urchin and the beautiful mother-of-pearl we know from the inside of shells we find on the beach are among the world's strongest materials, even though they consist of the same crystals found in soft, porous chalk. These materials are created by the animals controlling the chemistry both inside and outside their cells. They can decide what kinds of crystals will be formed and how these crystals will grow. The combination of strength and elasticity that we seek when assembling different materials such as steel and concrete is achieved by such organisms by combining organic molecules and inorganic crystals.

If we could recreate the materials already made by life on Earth, we'd be able to produce even better materials than the ones we have now—using only a fraction of the energy. This is a huge field of research today.

Even I myself am researching how we can use bacteria to develop limestone concrete. After all, limestone was made of living organisms in the first place. The concrete we're developing doesn't have the right chemistry to be reinforced with steel, but if we find suitable organic fibers to mix into the concrete, it may nevertheless be the material of the future in large constructions. I'm also trying to develop a concrete that can be crushed and dissolved when old buildings are demolished so that the sand and gravel can be used in fresh concrete for new structures.

This bacterial concrete is one of the possible alternatives that would allow us to build in the future without the energy use and carbon dioxide emissions we have today. We spend a lot of time in our project discussing how the choices we make can affect the environment and resource use in the future. Although we're a long way from our goal and the chance of success isn't all that high, I'm happy to be working on the future of civilization through my research.

6 | Multitalented Carbon:
Nails, Rubber, and Plastic

EVERY ONCE IN A WHILE, I stop by the hospital on the way to work. I get to drink as much apple juice as I want, sit in a comfortable chair, and have a thick needle stuck into the largest blood vessel in my arm. Giving blood is a quick and easy way to feel you're contributing to something important.

It also means a lot of disposable blood-drawing equipment. First, a little packet for disinfecting the skin. Then, there's also the syringe, the tubing the blood flows through, and four to five tubes for blood tests. After being used just once, everything disappears into a big bag that is sent into a hospital's incinerators.

It's a tremendous boon to the health-care system that such equipment can be used for a single task and then tossed into closed containers and incinerated. If all gloves, surgical masks, and diapers had to be reused, an enormous infrastructure for disinfection management would have to exist, including the considerable consumption of water, chemicals, and energy. This was how it was before we had access to the synthetic, oil-based plastics we use today. Glass syringes, steel kidney bowls, and cotton bandages all had to be sterilized between each and every use. Blood and intravenous fluids were transfused from glass bottles through rubber tubing. This rubber was also sterilized and reused, but it cracked easily and was difficult to keep clean with repeated use.

The storage of blood in glass jars with rubber corks (which incidentally were never completely airtight) made it cumbersome to store and distribute blood to the injured during World War II. The first real breakthrough for blood banks came in 1950, when soft, impermeable plastic bags for storing blood were developed. Now, blood could be stored easily and securely and transported between different blood banks as needed. This steady supply of blood for patients in need has been crucial in developing treatments for a wide range of diseases and injuries.

NATURAL RUBBER AND VENERABLE VULCANIZATION

Before oil-based plastics appeared on the scene, natural rubber was the most important material when you needed something that was durable, waterproof, and strong. Rubber is manufactured from the sap of a number of tropical trees in Africa, Asia, and South America. Today, most natural rubber is produced on plantations in Southeast Asia using a rubber tree originating from the Amazon.

The export of rubber from South America to North America and Europe began in earnest at the end of the 1700s, and the new product quickly found a variety of uses. As a textile coating, it was used to make raincoats, rubber hoses, and car tires. Glass jars sealed with rubber gaskets (like what we see with Kilner or Mason jars) were manufactured from the mid-1800s, enabling private individuals to preserve summer fruits and berries without refrigerators or freezers. Gutta-percha, a harder type of natural rubber from Southeast Asia, was made into bottle caps and golf balls. It was also ideal as insulation for electrical wiring and was instrumental in the implementation of the first submarine telegraph cable from Dover, England, to Calais, France, in 1851, and another across the Atlantic in 1866.

Rubber as a material is made up of large molecules, each of which can contain tens of thousands of carbon atoms. These atoms have a special ability to bond with each other in that two atoms can share either one, two, or three electrons from each—thereby creating one, two, or three shared electron pairs. Such variations allow carbon to appear in countless structures: from straight or branched chains to rings, sheets, and tubes. These carbon structures, together with oxygen, hydrogen, and small amounts of other elements, make up the majority of the content of our bodies and those of all other living beings. Collectively, they are called organic molecules.

Natural rubber is strong and elastic because it consists of long, coiled-up chains of carbon. As the material is stretched, each of these chains can straighten and slide past its neighbors, causing a lump of rubber to change shape. However, pure natural rubber is easily affected by temperature changes, which makes it unsuitable for many applications. When it's cold, the carbon chains become stiffer, making it difficult for them to crawl past their neighbors. This can make the material hard and brittle. When hot, natural rubber becomes soft and pliable, but it can also get sticky.

In the mid-1800s, chemists figured out how to improve natural rubber's properties: If they heated a mixture of rubber and sulfur, the rubber didn't melt as it would otherwise. Instead, it became harder, stretchier, and less affected by heat and cold. This process, which became known as vulcanization, led to an immense increase in rubber usage in a worldwide "rubber rush" toward the end of the nineteenth century. Now, rubber could be used in bicycle tires, and bike rides became much more comfortable for thousands of new cyclists. Car tires made of vulcanized rubber helped lay the foundations of the emerging market for private car ownership. Vulcanized rubber could also be used for pipes and gaskets, and provided good electrical insulation in telegraph, telephone, and power lines.

However, another consequence of the huge demand for rubber was that at least ten million people lost their lives in the Congo. At that time, the Congo Free State was privately controlled by King Leopold II of Belgium. He was able to make a profit on the sale of Congolese ivory, which was used to produce common objects such as knife handles, billiard balls, combs, piano keys, chess pieces, and snuff boxes in the 1800s. Still, his earnings weren't substantial enough to make up for the considerable debt he'd accrued by investing in the Congo.

When the rubber rush took off, most of the rubber on the market came from rubber tree plantations. Rubber trees needed several years to grow large enough to be tapped, but in the Congo, rubber could be harvested from vines that grew freely in the jungle. The only thing needed to produce large quantities for the world market was therefore labor. The Belgians implemented powerful measures to set the Congolese people to work. They would hold all women and children in a village hostage until the men had delivered a certain amount of rubber. Men, women, and children alike could have their hands or feet cut off as punishment for not meeting their rubber quota. Both outright murder and indirect suffering, such as hunger, exhaustion, and disease, led to the deaths of millions. The details of these atrocities rarely reached Westerners cycling around on their bouncy rubber bike tires.

In the beginning, no one understood why rubber gained such excellent properties when it was heated with sulfur. Today, chemists know that sulfur atoms attach themselves to the hot carbon molecules to form tiny sulfur bridges that span across individual carbon chains. This is how rubber goes from behaving like a handful of boiled spaghetti, where each thread is free to move in relation to the others, to becoming a rigid network of threads that are interconnected at many points—much like how a ball of yarn becomes a tightly knitted sweater. We find the same principle in our own bodies; our fingernails and toenails are made of keratin, which also consists of long carbon chains bound together in a crisscross pattern by sulfur atoms.

More sulfur leads to more connection points and, therefore, a harder, stiffer material. The vulcanization process for making rubber tires for bicycles and cars uses between 3 and 4 percent sulfur. If the sulfur content is increased to about one part sulfur and two parts rubber, the end product becomes incredibly hard. This material is called ebonite, because of its remarkable resemblance to ebony, and was ultimately used to mass-produce hard objects such as fountain pens and dentures.

FROM TIMBER TO TEXTILES

Trees, nature's skyscrapers, get their stiffness and strength from two large carbon molecules: cellulose and lignin. The cellulose is produced in the trees' inner machinery when many ring-shaped glucose molecules are connected through strong bonds. Whenever molecules are made up of repetitions of smaller units (such as the glucose in cellulose), they're called polymers. And there are *countless* natural polymers. We humans are full of them: our DNA, which makes up all of our genes; proteins, which give the body its structure and function; and the keratin in our nails are all polymers.

One cellulose molecule can contain several thousand units of glucose assembled into a long, straight chain. In trees, these long cellulose fibers provide strength and resistance to stress so the trees don't break in the wind. Cellulose is mixed with the branched polymer lignin, which keeps the cellulose molecules in place, much like how reinforcing iron is held in place by concrete in human-made skyscrapers. The combination of lignin and cellulose, together with the structure of the tree's plant cells, gives wood its specific properties. These properties are reliable enough for us to build houses and bridges, as well as furniture, tools, and paper.

Like natural rubber, cellulose should theoretically be a molecule that is well suited for making other materials. However, it's difficult to process cellulose by chemical methods. Cellulose doesn't dissolve in water, and when heated, it breaks down and turns into smoke. It was only when someone—perhaps even by accident—figured out that cellulose could react with nitric and sulfuric acid to form the combustible substance nitrocellulose that cellulose could be used for something other than its original fibers. Nitrocellulose could be made into a hard material, which was cast into molds to make objects such as buttons, combs, and knife handles. These had previously been made from expensive materials such as metal, horn, ivory, and turtle shells. One of the first major markets for nitrocellulose was for billiard balls, which until then had been made of ivory. Many elephants may have been spared as a result of the invention of this new material.

Another nitrocellulose-based material was called celluloid (also known as nitrate film) and used as film during the birth of cinema. One disadvantage of nitrate film, however, is that it is extremely combustible, and it naturally contains enough oxygen to burn even without any external oxygen supply, meaning that it can continue burning even under water. This led to several disastrous fires in movie theaters. Cinema technicians had to get special training in fire safety, and it was illegal to transport films on the London Underground until a safer form of cellulose film became universal around 1950. Today, celluloid is still used in some paints, varnishes, nail polishes, and explosives.

Later, chemical methods of dissolving cellulose in water were developed. When this solution is pressed out through a narrow nozzle, strong fibers called viscose are formed. Chemically, viscose bears many similarities to cotton, and the fibers can be woven into a comfortable fabric used in many clothes. When used in textiles, viscose is often called rayon. When the same aqueous solution is pressed out into a thin layer, it forms a thin, transparent film called cellophane. This film is resistant to moisture, oil, and bacteria, and therefore well suited for food storage.

PLASTICS OF THE PAST

Avocados, cucumbers, bell peppers, tomatoes, ground beef, frozen fish fillets, yogurt, Parmesan cheese, tortillas: Almost everything we buy at the grocery store is packaged in plastic. Sometimes the plastic is thick and soft, other times thin and almost brittle. Sometimes it even has indentations in it so that each individual avocado can rest in its own sturdy plastic cup, molded specifically to their shape and size. Plastic can protect against bruising and other damage, keep moisture inside and oxygen out, and act as a barrier against mildew, bacteria, and viruses. Without this packaging, it would be much more difficult to send food to my local shop from the countries where it was produced. Even though some of the plastic that ends up in my shopping bag is unnecessary, proper packaging can prevent food from going bad during transportation and needing to be thrown away when it reaches a store or the kitchen counter.

When it comes to thinking up the kinds of materials that can be made from the carbon molecules produced by living plants and animals, it's really only our imagination that limits us. However, the abundance of cheap plastic we surround ourselves with doesn't come from anything that grows on Earth today; the building blocks of my toothbrush came into existence millions of years ago.

Plants and animals that die don't always become soil. When organic matter ends up in bogs or at the bottom of lakes or the seabed, there often isn't enough oxygen present for microorganisms to break down all of the large carbon molecules. Over time, the remains of living beings will be buried beneath ever-thicker layers of dust, sand, and gravel and pushed down into Earth's crust. There, the pressure and temperature increase and the large carbon molecules start breaking down into smaller pieces. At about 2 miles (3 km) down, the molecules will have become so short that what had been solid starts to become liquid, and the remains of algae and other small creatures that lived in the sea can become oil. In addition, some small molecules are always formed when organic material is heated, and these become what is known as natural gas. The remains of large organisms such as dinosaurs and trees rarely reach this point; they turn into coal.

When humans began extracting oil and using it as energy, a number of chemists started experimenting with making plastic materials from the carbon compounds found in oil. In 1907, Belgian American Leo Baekeland became the first person to succeed at making plastic from these fossil raw materials. He did this in a custom-built laboratory in his own backyard and unashamedly named the material after himself, calling it "Bakelite." Bakelite was hard and moldable and could be used extensively, including for the insulation of electrical components and in car parts, telephones, and toothbrushes.

Gradually, an almost infinite variety of synthetic (that is to say, oil-based) plastic materials were developed. Common to all of these is that they can be made incredibly strong, they're much cheaper than any of their nature-based counterparts, and it's possible to make a lot of them. Cellulose polymers had already replaced natural polymers such as silk, ivory, and horn in some products, but now these synthetic materials took over almost entirely. Over the course of the 1900s, the use of plastic materials spread to molded plastics, plastic films, fibers, laminates, adhesives, and surface coatings.

Plastic materials don't necessarily consist solely of large carbon molecules. They often also contain other materials such as soot particles, lime, clay, or wood dust, all of which provide strength or other properties to the materials. It's also common to mix different types of fibers into plastics, and the principle is the same as when introducing reinforcing steel into concrete structures. A good example is the use of fiberglass with plastic materials such as polyester or epoxy, which is then used to cast strong, lightweight objects like boat hulls and wind turbine blades.

Today, nearly 400 million metric tons of plastic are produced each year. In comparison, the world's yearly oil consumption is over 4 billion metric tons. That is to say that plastic production makes up a tenth of the oil we pump out of the ground. At the same time, the amount of plastic we've manufactured over the last 150 years has begun to become significant—and not least conspicuous everywhere we look.

THE TRASH ISLAND

Out in the Pacific Ocean somewhere between New Zealand and Chile lies Henderson Island. This uninhabited paradise is on UNESCO's list of World Heritage sites, named as a place of outstanding universal value due to the fact that it is one of the few atolls in the world where the ecosystem has been able to evolve virtually undisturbed by humans.

Henderson Island is part of the Pitcairn archipelago. The closest inhabited island is Pitcairn, populated by about fifty descendents of the mutineers who took over the British ship The HMS *Bounty* in 1789. A couple of times a year, residents make the 70-mile (110 km) boat journey to Henderson to collect timber, but otherwise, the island is left to itself. The nearest populated area of considerable size is over 3,000 miles (5,000 km) away.

The Pacific isn't just an enormous body of water; it also has large and constant currents. These currents move water north along the coast of South America before bending west along the Equator, then southward, and then east again along Antarctica. Objects in the water are transported with these currents and often end up in the center of this large rotational pattern. Henderson Island is located on the outskirts of large, rotating water masses that serve as a gathering place for everything that has been lost and forgotten.

We know there's a lot of garbage in the ocean, but it's hard to determine exactly how much. In 2015, a group of researchers traveled to Henderson Island to count pieces of garbage, choosing the island because it is uninhabited and so rarely visited. All the garbage on the beaches therefore must have come from the ocean, and since no one had been there to clean it up, the amount of garbage would reflect everything that has ever been washed up on the island.

After the researchers had collected and counted garbage for nearly three months, they had their answer: The beaches on the 5-mile by 3-mile (8 km by 5 km) island were covered with 37.7 million pieces of trash. The garbage was so dense that it made it difficult for turtles to build their nesting pits and for hatchlings to get out to the sea. Hermit crabs were using cans as shells. The total weight of the garbage was estimated to be 17.6 metric tons, and every day, a few hundred new pieces of trash were washed up on the beaches.

Only two thousandths of the debris found on Henderson Island were made of materials other than plastic. Metal sinks; plastic floats. Wood and paper are broken down by microorganisms, but even though plastics are made of the same molecules, the chemical processes that make natural polymers into these strong and durable materials also make it difficult for life on Earth to break plastic down into its components. The plastic we make could exist on Earth for hundreds or even thousands of years.

WHAT DO WE DO WITH ALL THIS PLASTIC?

It's hard to avoid plastic ending up in nature. I often see it in my hometown of Oslo—for example, in public trash cans, where people toss their yogurt cups and burger wrappers, crows drag the garbage out to get at the last remnants of food. The packaging is then simply left lying on the ground. Plastic also ends up in nature through agriculture, where the plastic around hay bales isn't always collected properly and sent to be recycled, and from fishing and fish farming, where gear is lost at sea. On Henderson Island, 6 percent of the plastic found there came from fishing, while as much as 11 percent was plastic granules (nurdles)—small plastic grains carried on ships from factories that make plastic from oil to the production plants that make plastic bags and garden furniture.

The plastic that isn't washed up on beaches and buried in the sand can be eaten by birds, mammals, and fish. In the winter of 2017, a whale with over forty plastic bags in its stomach washed up on a Norwegian beach. It was so sick that it had to be euthanized. Much of the plastic in nature gets torn into smaller and smaller pieces that can be eaten by the tiniest animals in the sea and move up the food chain until it ends up on our plates. This is how our own waste can end up in our own bodies, blood, and cells.

Of course, it would be best if we could simply collect all the used plastic and recycle it. Much of the Western world has systems in place for collecting waste plastics from households. In Oslo, we dispose of our plastic waste in blue bags, which are then shipped to Germany to be sorted into different types, some of which is melted down into new plastic, but some of which is converted into chemicals. Unfortunately, plastic is not very well suited for recycling; when the long polymer chains are bundled together in networks, it becomes impossible to tear them apart and return them to their original components. It is, however, possible to at least melt and reuse some of the plastic where the polymers aren't chemically linked to one another, but the large molecules will often break down at high temperatures. It's also a problem that there are so many types of polymers in the plastic we send for recycling. These different polymers have different properties that can't be mixed together to make a usable material. When plastic is recycled today, it's largely broken down into shorter and shorter molecules until it's suitable for nothing but combustion.

Burning used plastic isn't necessarily a bad solution, however. The carbon compounds that make up plastic contain a lot of energy, just like the oil it was made from. When we use part of the oil to make plastic and then burn it, we are in a way actually benefiting more from the oil than if we had just burned it in the first place. Of course, we do want to stop burning oil to limit fossil carbon emissions into the atmosphere, but the amount of oil we burn directly is much, much more than the amount we use for plastics and that we will probably end up burning eventually anyway.

A bigger problem is that plastic doesn't burn cleanly. We've all been warned against burning plastic in the fireplace. At low temperatures, it can break down into medium-sized molecules that can harm both us and nature. To burn plastics safely, it must be done in high-temperature industrial furnaces with safe and effective filters to avoid emissions.

Degradable plastic has been introduced as another solution to the waste problem. Plastics that can be broken down by microorganisms can be made from both fossil raw materials and renewable materials such as cellulose. The challenge is making a plastic material that has the properties we want—strength, durability, and imperviousness to water and oxygen—and that can also be broken down by microorganisms after being used. Additionally, some of the degradable plastic materials we find on the market today are downright scary. They can be broken down until we no longer see them with the naked eye, but microscopically small plastic particles known as microplastics still remain in the soil. For degradable plastics to be safe for the environment, polymers must be able to be broken down into small molecules that are already naturally found in soil and water.

Degradable plastic isn't made for recycling but rather for use and disposal. Once used, it doesn't give us new materials or energy. Plastic that can be eaten by microorganisms should be reserved for cases where we know for sure that we won't be able to collect all the used plastic. In other applications, such as computers and mobile phones, it's more important that the plastic used in production can be recycled or reused.

PLASTIC AFTER OIL

Plastic is everywhere. I have plastic in my toothpaste tube, kitchen utensils, furniture, clothes, bike, car, skis, phone, and computer. Before we started using oil-based plastics, millions of turtles, elephants, and humans were killed to meet our needs for polymer materials. At that time, there were around 1.5 billion people on Earth. Today there are more than 7 billion of us, and it's been estimated that plastic production will increase to 1 billion metric tons annually by the end of this century. Producing this much plastic will require a quarter of today's annual oil production. What would happen if we had to stop making plastic from fossil sources?

We've already made plastic products from cellulose and other natural polymers. The first Legos were made from cellulose, for example. Today, there are methods for making single fibers of cellulose, which can provide extremely strong materials in combination with other polymers. Similarly strong fibers can be made from chitin, which is found in the shells of shrimp and crabs. Researchers are also working on making plastic products from lignin (the other large molecule we get from wood) and by processing polymers from plant oils. These raw materials are far more abundant in nature than natural rubber and gutta-percha. In theory, it should be possible to produce all the products we need from natural raw materials.

We can also use microorganisms to make new polymers for us. We're already able to buy plastics made from lactic acid, which is produced by bacteria or fungi that eat sugar and starch. Other bacteria make excellent cellulose fibers. When researchers have studied life-forms that live under extreme conditions such as inside volcanoes or in the deep sea, they've found organisms that are robust enough to function well in industrial processes. New advances in biotechnology have given us the opportunity to manipulate the genes of bacteria and fungi so they can produce more of what we want or even create new materials.

Now, the challenge for scientists and industry lies in exploring all possibilities for producing materials from renewable polymer sources, as has been the case over the past century for petroleum products. At the same time, we need to figure out how to get enough plant-based raw materials to produce the plastic we need without overloading ecosystems and without it happening at the expense of food production. If we manage to do all of these things, our descendants might also be able to enjoy all of the benefits plastics offer us today.

7 | Potassium, Nitrogen, and Phosphorus: The Elements That Give Us Food

THE FACT THAT WE USE elements from our planet to build the objects we surround ourselves with may seem like a given. But we probably think a lot less about the fact that we need elements from rock, water, and air to build ourselves. We do this by extracting these elements and turning them into artificial fertilizers that become a part of plants and later our bodies when we eat those plants, or when we eat animals that ate those plants. The most common fertilizer on the market is called NPK, after the chemical names of the elements that are the most important for the plants to be supplied with: nitrogen (N), phosphorous (P), and potassium (K).

Most of the other elements we use in our civilization, such as iron for tools, copper for power lines, and gold as a means of payment, can be exchanged with others if need be. Food production is a different story, however. Without nitrogen, phosphorus, and potassium, we can't exist.

THE JOURNEY TO THE DEAD SEA

In the fall of 2016, I went to Israel for the first time for a meeting of a European research project in which Israel was a partner. To gather the researchers in a place where we could work closely together for one week, undisturbed by everything going on in a city or a university environment, our Israeli partner had decided that the meeting should be held at a hotel on the Dead Sea.

Since I'd never been to the area before, I took a look at Google Maps before we left to check out where we were going. The region looks strange, to put it lightly. The Dead Sea is split in two by a strip of land, making one lake in the north and one in the south. The northern part looks like an ordinary lake surrounded by a desert. The lake in the south, on the other hand, looks like nothing I'd ever seen; it's divided into areas delimited by straight lines, much like plots of land. The water between these lines is colored in shades ranging from turquoise to dark blue, and the border between Israel and Jordan that runs through the middle of the lake is marked by a thick strip of land. I became quite curious about what these patterns might mean.

After spending a few days sightseeing around Tel Aviv, I set off on a bus journey to the Dead Sea, which took us through a dry, flat desert until we were at the edge of a cliff. From here, the road took a steep turn down toward a lake that looked just like a blue and turquoise patchwork quilt—exactly as it had appeared in the pictures I'd seen. There was still a good way down when we passed a sign informing us we were at sea level. When we finally reached the bottom of the valley, we passed an industrial plant with large storage tanks, conveyor belts, and pipes. The straight lines in the lake turned out to be long levees with giant machines running back and forth between the patches of blue-green water.

It's true that the southern part of the Dead Sea is different from the north. It used to be one large connected lake, where the surface level of about 1,300 feet (roughly 400 m) below sea level was maintained by the fact that there was always as much water entering into the lake as was evaporating. This changed in the 1960s and '70s when Israel, followed by Jordan and Syria, built pumping facilities and started diverting the water that used to flow into the Dead Sea into pipelines for use in agriculture and in the cities. Today, the surface of the Dead Sea is more than 130 feet (40 m) lower than before the pumping plants were built. Recently, the water level has been dropping by about 3 feet (1 m) per year. The northern and southern parts of the lake have now completely separated from one another at a point where there used to be a cape.

The southern part of the lake, which in most places maxes out at just a few meters deep, has been converted into a huge mineral extraction facility. Because of natural evaporation, the water in the Dead Sea is already much saltier than other oceans and lakes on Earth. When the water is moved into shallow pools, evaporation speeds up—just as laundry dries faster if you hang it up and don't leave it in a wet pile. Evaporation can only occur on the surface of the water where the sun heats it up and causes it to disappear into the air as water vapor. As the amount of water decreases, the salt that was dissolved in it starts to precipitate as crystals. First, sodium chloride (common table salt) is created, and then calcium chloride (often used as road salt). These crystals sink and form a salt crust at the bottom of the pond. The water is then pumped onto the next evaporation dam, where crystals of the coveted mineral carnallite are formed, scraped up, and transported to the processing plant at the edge of the lake. Carnallite is valuable because it contains potassium, which is used in fertilizers because plants need it to live—just as we humans do.

NUTRIENTS IN OUR NERVES

Potassium is an element that dissolves easily in water. It does so by emitting one electron (which it's eager to get rid of) to another atom found in the water. By doing this, potassium can float around with its positive charge, surrounded by a nice bunch of water molecules. In the human body, potassium plays an important role in the electrical signals transmitted through our nerve pathways. When I looked out over the Dead Sea, the light that hit my eyes set off a reaction in which tiny gateways in the optic nerve allowed potassium to flow in and out of the nerve cells. This impulse was transmitted through the optic nerve and all the way into my brain, where similar potassium signals were sent at lightning speed between my brain cells, thus storing the image of the Dead Sea in my memory.

Since potassium is water soluble, we're constantly losing some of it from our bodies when we sweat, pee, or cry. The only way we can get it back into our bodies is by eating plants that contain potassium or animals that have eaten plants that contain potassium. The element makes its way into plants through their roots when they absorb the water sitting between particles of the soil where they're growing. If this water doesn't contain potassium, then the plants aren't able to grow as they should.

When plants and animals die and remain in the ground, they get broken down by small and large living organisms. The roots of living plants then gradually absorb the nutrients that were found in the dead material. In this way, potassium can circulate from one living being to another indefinitely. However, if we look at a specific area such as a forest, this isn't entirely accurate; after all, it is inevitable that some nutrients will disappear from a given area. For example, an animal can eat a plant in the forest but die somewhere else. Leaves from the trees can be blown away in the wind. Soil and remnants of plants and animals can be transported out into the sea by streams and rivers along with the nutrients they contained.

Over a long period of time, this loss of potassium from the forest could cause life there to dwindle. Fortunately, there is another source of potassium: the rock the forest is growing on. Bedrock contains many of the nutrients that life requires. Perhaps this rock was once the bottom of an ocean—one that was strewn with microscopic remains of another forest long, long ago. When this rock weathers, it breaks apart, and different kinds of fungi and bacteria can work on the surfaces of the ever-smaller mineral particles to release the nutrients needed to sustain life up above. Loss is inevitable, but as long as the losses aren't greater than the supply of nutrients coming from the weathering of rock, life in the area really can continue into eternity.

The situation is quite different where we humans are engaging in farming, however. The whole point of cultivating grain in a field is to let the plants absorb energy from the sun, carbon from the air, and nutrients from the soil—and then transport this away from the field to the places where people are living. In cultivated soil, the release of nutrients through bedrock weathering is far too slow to replace what is constantly being removed from it. We can restore the soil's nutrients by composting plants and by fertilizing the soil with animal dung, but it's still difficult to replenish enough to make up for everything we take away when we produce our food.

What we can do to maintain the soil's ability to produce food is to lend nature a hand by making nutrients from air, water, and mountains more readily available to the plants. This is what we call chemical fertilization. By taking control of how this nourishment is supplied to the soil, we humans have freed ourselves from a fundamental constraint that applies to all other creatures on the planet. But is this really a solution for the future? Is there a chance that we might run out of some of these elements?

POTASSIUM FROM WATER

Potassium is everywhere around us. It can be found dissolved in all water on Earth's surface, including rainwater. Salty seawater contains much more potassium than rainwater or freshwater in rivers and lakes, but the concentration of potassium in the ocean is too low for it to be worthwhile to try to extract it directly. To make chemical fertilizer, we need sources where potassium is extra concentrated.

Today, potassium is extracted from areas where seawater has been evaporating over thousands and millions of years. Most of these deposits can be found in the ground as thick layers of salt, the remnants of earlier salt lakes. If the salt layers are near the surface, they can be excavated as in other mining operations. However, many of the deposits are quite deep, and when machines have to dig more than half a mile into Earth's crust, mining efforts can become quite costly. Since potassium dissolves so easily in water, this challenge is solved by pumping water down into the deep salt layers. When the water comes back to the surface, it's carrying salts that can be deposited into shallow evaporation pools. From here, the extraction takes place in the same way as in the Dead Sea. In satellite images, such potassium mines look like industrial plants surrounded by blue and turquoise ponds—beautiful, but dead. Canada is the world's largest potassium producer, followed by Russia, Belarus, and China. Only 5 percent of the potassium extracted is used for purposes other than fertilizer.

Although the reserves stated in geological reports indicate there's only enough potassium for about a hundred more years (at today's rate of consumption), estimated resources are actually large enough to produce potassium for thousands of years. However, most of these are quite deep beneath Earth's surface. The challenge for the future isn't finding enough potassium but having enough energy and water to be able to extract what's available.

Some countries are lucky enough to have an abundance of clean water. In many parts of the world, however, water is a scarce resource. When I was visiting the Dead Sea, I could see how the flowers, trees, and lawns around the hotels got their water from a network of pipes supplied with industrial wastewater: one hole for each flower, and large signs with skulls to discourage thirsty tourists from taking a sip.

We need clean water for drinking and food production, but also for producing fertilizer and extracting metals and other raw materials from the earth. Nature produces clean water for us in two ways. One is that the sun causes water to evaporate from the oceans (or from the evaporation pools in potassium mines). The water vapor is transported through the air currents over land, where it falls as rain and collects in streams and rivers, ready for use. In addition, nature has its own filter to purify dirty water. When water flows through the ground, bacteria and other microorganisms break down substances that had been dissolved in the water. Other substances will adhere to the sand or clay as the water flows past.

Today, we don't just use the water that nature purifies for us each day; we also use water that was cleaned thousands of years ago and stored deep down in the ground in the form of groundwater. Several of these old aquifers are being emptied because the refilling of rainwater from the surface is far too slow to make up for how much we're draining them. In some places, there is such a scarcity of natural sources of clean water that it's become necessary to purify seawater of salt, but this requires a great deal of energy. The needs we have for water, energy, and fertilizer are closely linked.

NITROGEN FROM AIR

The first element in NPK fertilizer, nitrogen, makes up about 3.2 percent of our body weight and plays a key role in the large molecules that build up our skin, hair, muscle fibers, tendons, and cartilage. It's also part of the molecules that control all of the chemical processes in our bodies. Without nitrogen, it would be impossible to make a human body that can function.

Most of the air we breathe (78 percent, in fact) consists of nitrogen, so you might think it's something we always have enough of. The problem is that nitrogen gas in the atmosphere consists of two nitrogen atoms that are strongly bonded. We simply breathe nitrogen in and back out again because our bodies aren't capable of tearing the nitrogen atoms from each other and making use of them. We therefore have to get the nitrogen we need through the food we eat.

Fortunately for life on Earth, there are bacteria that manage to break the bond in nitrogen molecules. These bacteria bind the released nitrogen atoms with three atoms of either hydrogen or oxygen, and when these compounds are dissolved in water, they can be absorbed and utilized by plants. Some plants, such as clovers, even allow such bacteria to reside in special tubers in their roots. The clover ensures that the bacteria is safe and sound, and in return receives a stable supply of nitrogen for its own growth. When the clover plant dies, the nitrogen stored in the plant can then be used by other plants growing nearby.

Nitrogen is an element that thrives well in gaseous form, which is why it's easy for it to be lost when dead plants and animals are broken down. You might recognize the pungent scent of ammonia ubiquitous in Porta Potties and barns. Ammonia consists of nitrogen and hydrogen and can be formed when organic material decomposes. This flow of nitrogen into the atmosphere makes the existence of organisms in the soil particularly vital, since they can do the job of extracting nitrogen from the air again.

The only natural process that makes nitrogen from the atmosphere accessible for plants without involving microorganisms is a lightning strike. When lightning strikes, so much energy is released that nitrogen and oxygen in the air can react with each other to form new molecules. In the early 1900s, Norwegian physicist Kristian Birkeland and engineer Sam Eyde figured out that they could mimic this process by using electricity to create artificial lightning in a laboratory. This was the first time anyone had managed to outsmart biological processes and produce nitrogen fertilizers directly from the air.

The Birkeland-Eyde process allowed Norwegian energy company Norsk Hydro to produce fertilizers with energy from hydropower. Norsk Hydro's fertilizer production was in many ways revolutionary and helped lay the groundwork for an enormous increase in world food production throughout the twentieth century. Nevertheless, it was soon replaced by the cheaper Haber-Bosch process, which is based on so-called natural gas—that is, gas from fossil sources.

In the Haber-Bosch process, natural gas gives us not only the energy required to break down the strong bond between nitrogen atoms but also the hydrogen atoms that the liberated nitrogen atoms can bind to. Completely pure carbon dioxide is produced as a by-product and sold both for beer production and water treatment plants. Today, over half of the nitrogen absorbed by the plants we grow from agriculture around the world comes from fertilizers. We've become addicted to building our bodies with nitrogen produced using industrial methods. How long can we continue this way?

If all of today's known natural gas reserves were to be used solely for fertilizer production, it would give us enough nitrogen fertilizer for about a thousand years before we ran out. Now, there's most likely more natural gas than what's known to be in reserves today. At the same time, it's unrealistic to imagine that natural gas won't be used for anything other than producing fertilizer. As fossil oil and gas deposits become smaller and more expensive, natural gas will become an even more sought-after raw material for a variety of chemical processes. We'll therefore eventually have to produce nitrogen fertilizers from something other than natural gas well before a thousand years have passed.

Even today, we're already working hard on developing alternative methods of production. One strategy is using solar energy, both for splitting nitrogen and for splitting water to produce hydrogen that can then react with the nitrogen. Others are working on developing the old Birkeland-Eyde process to make it more energy efficient. It might even be just a few years before farmers can make their own nitrogen fertilizer using the solar cells they have on their roofs.

When bacteria capture nitrogen, they do so by creating organic molecules that cause the nitrogen gas to react with hydrogen without requiring much energy. The molecules somehow manage to "coax" the atoms into place. Now that we humans have started to develop some useful tools for editing the genes of both bacteria and plants, we have the opportunity to change agriculture-grown plants to better enable them to capture nitrogen from the air on their own, or to enter into new collaborations with customized bacteria just as clovers do. These genetically modified plants would in theory make it possible to grow all of the food we need without even needing nitrogen fertilizers. The issue around nitrogen for food production is a technical challenge, but one with a number of possible solutions. In other words: We won't run out of nitrogen.

PHOSPHORUS FROM ROCKS

The last of the vital elements, phosphorus, isn't found in the atmosphere. We don't find much of it in water, either, since it has the tendency to stick to mineral surfaces and thrives better in solid form than dissolved in water. We therefore need to look to solid rock to harvest phosphorus.

Phosphorus accounts for about 1 percent of an adult human's body weight, and most of this is found in the skeleton. However, it also plays a crucial role in organisms without bone structures. The recipe for my body—who I am—is written inside each of my cells with chemical writing. The chemical alphabet consists of only four letters, and these letters constitute the steps of a long molecule that looks like a twisted ladder. Phosphorus atoms keep the steps in the ladder connected. Without phosphorus, there would be no DNA—and therefore no life.

Phosphorus sources such as human urine and bone meal from domesticated animals have been used as fertilizer for centuries all across the world, but in the mid-1800s, farmers also started using phosphorus from geological deposits. The first lucrative source was bird dung, also called guano, which can be found in vast quantities on some islands where seabirds have been breeding for thousands of years. These deposits of "fossil" bird dung were extracted on a large scale, but since the deposits were limited, it didn't take long before more phosphorus from rock was being produced than from guano.

Since the 1960s, soils have been supplied with more phosphorus from geological sources than from livestock and plant remains, and today, over three times as much geological phosphorus is supplied as is recycled in these biological processes. If we were to stop producing phosphorus from rock today, we'd have to reduce food production to a quarter of today's levels. Geological phosphorus is even added in organic farming, but usually as ground rock. For conventional agriculture, the rock is often treated in a variety of chemical processes to make the phosphorus as readily available to the plants as possible.

A handful of countries dominate today's phosphorus sources. The largest and most important are Morocco, the United States, and China. Morocco alone controls over two thirds of the world's known phosphorus reserves. A large proportion of these are located in the disputed area of Western Sahara, which would be sitting on the world's second largest phosphorus reserve if they'd achieved independence from Morocco. Many are concerned that Morocco will have close to a monopoly on phosphorus in the future, thus in a way gaining control over the world's food production.

In some places on the seabed, there are sediments that are so rich in phosphorus that it could be profitable to extract them. Large deposits of this type can, for example, be found off the coast of New Zealand and Namibia. To extract phosphorus from the seabed, the top layer of sediment must be sucked up into a boat, where the phosphorus particles are then sorted out before the rest of the material is pumped back to the seabed. This destroys life on the seabed during the extraction period, and even though supporters of this process believe that life can return quite quickly, uncertainty around how damaging this can be for sea life in the long run has prevented such projects from starting up so far.

The documented phosphorus reserves are large enough to last for over 300 years with today's use. However, phosphorus consumption is increasing and expected to continue to do so alongside population growth in the coming years. Some scientists have warned that the world will face a dramatic shortage of phosphorus for food production in less than 100 years, and that it will only be a few decades before we start noticing this in the form of rising food prices worldwide. Others believe we can extract phosphorus for over 1,100 more years at today's consumption levels—that is, if we also count on there being deposits that haven't yet been mapped.

Only 20 percent of the phosphorus extracted in the world's mines finds its way to the food we eat; the rest is lost somewhere along the way. Some is already lost in the mines themselves. Some disappears in the chemical treatment of the phosphate rock and in the transport and distribution of fertilizer. Some is lost when crops fail due to plant diseases, storms, or fires. Of the phosphorus that actually ends up in human food, a third is never even eaten. Since the phosphorus we supply to our soil will always stick to the particles the soil is made up of, the greatest loss of phosphorus—about half of what we extract—occurs through soil loss.

The cultivable soil on our planet has been building up ever since the first living organisms emerged from the sea. In order for topsoil to be made, the rock it lies on must be broken up and disintegrated so that nutrients can be released. Small and large organisms such as bacteria, fungi, and earthworms ensure that the nutrients from the rock are mixed with organic matter from dead plants and animals. In total, nature spends about a hundred years on making a 1-inch (2.5 cm) thick layer of topsoil.

Erosion can cause this precious soil to disappear from our agricultural areas and into the sea. In a newly plowed field, the soil particles are lying unprotected against wind and weather, and a heavy rainfall can easily turn rivers brown with lost topsoil. After a drought, wind can swirl the top layer of soil into enormous dust clouds. This happened on the cultivated American prairies during the great disaster known as the Dust Bowl in the 1930s, when tens of thousands of families were forced to leave their homes and farms. In this area, as much as half of the topsoil has been lost since Europeans first started farming there. In the Dead Sea area, falling sea levels are causing the fertile soil to be washed down the slopes every time it rains.

Today, the world's topsoil is disappearing between ten and one hundred times faster than new soil can be formed through natural processes. There are good strategies for limiting erosion from agricultural land, such as plowing less or using plants that cover the soil as much as possible. On the other hand, we must expect that climate change will lead to more and more storms and floods, which increase erosion. If the loss of topsoil continues at today's speed, we'll have serious problems growing enough food for the world's population in the future, no matter how large the phosphorus deposits.

There will come a day when it will be too costly to extract phosphorus from geological sources to supply our soil. To continue our existence into the future, we must reach a point where we're not losing more phosphorus than what nature supplies us with when the bedrock weathers. This would mean a major reduction from today's phosphorus loss, which is six times greater than the natural supply. There is the possibility of optimizing agriculture to get as much phosphorus as possible from the surrounding environment, for example by letting livestock graze on uncultivated land, thus bringing phosphorus "back home to the farm." However, that still won't make up for more than about a third of today's losses.

In order for everyone to have enough food in the future, we're either going to have to reduce Earth's population considerably (a rather grim prospect) or make sure we're less dependent on geological phosphorus before it's too late. We can do this by eating less meat, preventing soil erosion, slowing down climate change, and recycling phosphorus from all parts of the chain—from dung heaps on the farm and food waste in the kitchen to sewage from our homes. In the future, we may need specialized toilets with an extra hole in the front that separates urine from feces, as has already been the case in some municipalities in Sweden. Not mixing urine and feces is a tactic that makes the road from sewage to fertilizer considerably easier.

NUTRIENTS GONE ASTRAY

The reason we need to supply artificial fertilizer to our fields in order to grow the food we eat today is because we remove much more nourishment from the soil than we give back through our own biological processes. But elements don't just disappear. The nitrogen, phosphorus, and potassium in the food we eat becomes part of our body for a while before being excreted through urine and feces. There are also nutrients in the parts of plants and animals that don't turn into food.

Today, only a small proportion of this nourishment is brought back to the land it came from. Transport costs are too high, and there's the risk that dangerous diseases can spread when human and animal feces are used as fertilizer. Instead, the nutrients end up where they shouldn't be. In rivers, lakes, and seas, large amounts of nitrogen and phosphorus can lead to enormous algae blooms, and this algae uses up all of the oxygen in the water, suffocating the fish that live deeper down. If we're able to develop effective methods for gathering nutrients and bringing them back to agricultural areas (without spreading disease), we can continue producing food in these areas while simultaneously keeping the seas and lakes in better condition for producing the fish we want to eat.

Our nutritious waste isn't the only thing the ecosystems in oceans and lakes are plagued by; they're also at the receiving end of quantities of the fertilizer we spread on our soil. When a farmer adds more than what the roots can handle, the surplus flows into rivers and streams. It can be hard to know exactly how much fertilizer a plant needs at a given time. Fortunately, this is a field in which great technological advances are being made. Computers can analyze images taken by drones to determine if the plants are missing anything. Then, the farmer can use computerized agricultural machinery to supply fertilizer as needed in the right place, and at the right time.

THE FUTURE OF THE DEAD SEA

The Dead Sea is a strange place.

At the northern end of the lake, the surface has sunk by several meters since the large spa hotels were built. Now, tourists must be driven down to the shore in small buses. The roads between the hotels are constantly under repair because the rainwater that flushes through what used to be a salty sea floor dissolves the salt and leaves cavities that collapse into gaping craters in the surface.

At the hotels on the south end, guests have to climb up stairs to get from the pool area to the beach. Construction work is constantly in progress. All of the salt that remains in the evaporation basins after the valuable minerals have been extracted causes the seabed to rise by roughly 7 inches (18 cm) a year. Water no longer flows from north to south; it has to be pumped up.

In a way, I find it heartbreaking to look at this poisonous, dead sea, where sparse shrubs and trash along the shoreline are slowly being encapsulated in a thick layer of salt. On the other hand, there's something majestic about the view. We humans have taken control over this entire ecosystem, this entire lake, and we use it to produce what we need most: the food that gives us the building blocks for life. Without fertilizer, we wouldn't be as many people on Earth as we are today. It's just that simple. And without the population explosion we've seen in recent years, we might not have had the technology and research we now have. I never would have traveled to Israel to meet scientists from all over the world. Potassium from the Dead Sea or a mine in Canada would never have flowed in and out of openings in my optic nerve to provide me with that striking image of the Israeli desert. I probably wouldn't have ever even existed.

8 | Without Energy, Nothing Happens

CIVILIZATION IS BUILT FROM PLASTIC, concrete, and metal. Every day, we crush up parts of Earth's crust so we can produce everything we need to maintain our way of life. And all of this activity is dependent on our access to another resource, one that isn't an element but the very basis for our being able to have a civilization in the first place: energy.

Nothing can happen without energy.

Without energy, the world would be cold and stagnant.

The light in the computer screen I look at, the electrical signals that are sent from the keys I press into the circuit board built of plastic, silicon, and metals—the processes that allow me to write this book—all of this is driven by the chemical energy of a lithium battery. I charged the battery earlier today by plugging my laptop into a wall outlet. My house is attached to copper wires that extend across all of Norway until they reach a power plant at the end of long water pipes by a dry riverbed. The water that once streamed down the mountainside is now passed through a turbine, and the movement of the turbine is transmitted to the electrons in the copper wire that move and push each other all the way into my battery.

ENERGY FROM THE SUN

In a way, my computer is therefore powered by water. In reality, though, it's really powered by the sun. The sun shone on the seas and caused water to evaporate and rise into the atmosphere. Later, this water fell as rain on the terrain surrounding the power plant's reservoir. The energy from the sun was transferred to the water molecules, and when the water ran through the turbine, the solar energy was transformed into electrical energy that I'm using to write these words at this very moment.

As I move my fingers to type, shift my eyes to look at the letters that appear on the screen, and think about which word should come next, I'm also using energy. All of the tasks my body performs are driven by the energy released from the food I've eaten, inside my cells.

I run on food, but I also really run on the sun. Each of the chemical bonds that are broken in my cells contains a tiny bit of solar energy. Plant photosynthesis is what captures all of the solar energy that the rest of us living beings on Earth use as fuel in our lives. Without photosynthesis, sunlight would have heated the ground, evaporated water, sped up the wind, and eventually just radiated back into space. Plants make sure solar energy takes a detour through the food chains before it disappears out into the universe once more.

Unlike other creatures here on Earth, we humans aren't satisfied with just releasing solar energy from photosynthesis within our own bodies by eating. A bonfire of crackling logs provides warmth to frozen hands. When food is cooked on the fire, it takes less energy to digest it. By using solar energy in timber to produce iron, people were given powerful tools that made their work more efficient and their crops grow larger. The extra food wasn't just used to feed a growing population; it was also used as animal feed. Then, the solar energy in grass and grain was converted into muscle work in oxen and horses so that people could do more and extract more.

DRAINING EARTH'S ENERGY STORES

Not all the solar energy captured by plants is released into the food chain. Living plants, animals, and fungi in the forests and soil contain a great deal of energy. When dead organisms don't decompose but instead accumulate in ever-thicker layers of soil, in marshes, and at the bottom of lakes and oceans, the stores of solar energy in the soil surface increases.

As the population on Earth continued to grow and each individual used more and more energy to make tools and build houses and roads, we started to drain this energy supply faster than it could be replenished. When Earth's population passed 1.5 billion at the turn of the last century, the solar energy stored in the living surface of Earth had been reduced by one-third from what it was around the time of Jesus's birth. Today, about half remains. The stores are reduced both because we humans use photosynthetic energy for food, wood, and fuel, and because our actions—which lead to things such as human-caused erosion and deforestation— make plants unable to capture as much solar energy as before.

The world's population passed 7 billion in 2012 and 7.7 billion in December of 2019. My family of five has access to about the same amount of energy as a landowner in ancient Rome with 3,000 slaves, or one in England in the nineteenth century with 1,500 workers and 200 strong horses. Our civilization is driven by an amount of energy that corresponds to about a quarter of all the energy the plants on Earth capture from the sun—energy that also needs to be used to power all of the ecosystems on the planet. The reason we're able to use so much energy while the planet can still be green and full of life is because we've stopped limiting ourselves to the energy the sun sends us each day. Over the last 150 years, we've become experts in exploiting multimillion-year-old solar energy; 85 percent of the energy that flows through civilization today comes from fossil energy sources.

THE SOCIETY WE WANT

In early agricultural society, it was important that almost every citizen did their turn in the fields—otherwise there wouldn't have been enough food to go around. The plants that were cultivated captured energy from the sun, and when people ate these plants, they received energy they could then use to grow *new* plants. The modest amount of energy to spare helped sustain a small ruling class of warriors, priests, and other government officials.

The development of tools allowed for less effort from each individual's muscles in order to provide for the population. Then, more of the inhabitants could use their energy for things other than agricultural work. Some became specialists in forging iron and making weapons and tools. When these tools were used in agriculture, they bestowed more food to the population than if the smith had spent his time digging in the soil with his own hands. In this way, one less person in the fields provided more food overall. The surplus of energy left room for more specialists, which also led to the development of new methods for getting as much useful work as possible out of the available energy.

Today, we've become so efficient that only a small percentage of us work in food production. Specialization and an energy surplus have laid the groundwork for an incredibly complex society with all of the advanced features we take for granted today—such as the internet, strawberries in the middle of winter, flights to the other side of the planet, and heart transplants.

Nothing happens without energy. The most important task in any society is therefore extracting energy and turning it into a form that can be used to do work—for fuel, say, or electricity. The next—not first—priority is growing food, since cultivation also requires energy.

The energy left over after society has secured the energy supply, produced food, and used that food to feed the population can then be used for education. The transfer of knowledge is necessary to maintain society's technological levels such that the energy surplus doesn't start to decline. Education also lays the foundation for further technology development and the potential for even greater energy surpluses in the future.

Next on the list of priorities are health services. We want society to take care of its members, even when they can no longer do work. The large surplus of energy we have today enables us to allow many of our citizens to work as doctors and nurses, not to mention run hospitals with advanced machines and research treatments for more and more diseases.

If we ask ourselves what makes life worth living, many will point out art and culture in a broad sense. We don't just want to work to survive. We also want to play sports, watch movies, go to plays, and sing in choirs. Today, we have plenty of surplus energy for this as well.

ENERGY IN, ENERGY OUT

The key to maintaining a complex society rich in culture and health services therefore requires having as much extra energy as possible once the fundamental needs have been covered. The most fundamental task is extracting energy and converting it into a usable form. But extracting energy also *requires* energy. This energy is used to, for example, drill a borehole, build an oil platform, and then pump up, transport, and refine crude oil that can be used as fuel in cars, boats, and tractors. The principle is the same as buying shares: You have to invest money up front, but you'll hopefully get more money back than you started with.

Researchers have estimated that in order to maintain what we consider to be a good way of life, we need energy sources that give back more than twenty times the energy used in its extraction. If the surplus drops to less than ten times the size of the "deposit," maintaining an industrial society can become challenging. A surplus of three times the deposit is estimated to be a kind of minimum measure of what the most primitive civilizations need to function. Before humans started agriculture, they regained about ten times the energy they put into hunting and gathering.

The oil that was mined to be used as energy in the 1930s was easy to extract. It practically only took sticking a pipe into the ground for the oil to come spilling out. Only one hundredth of the energy in the oil that was collected had to be used to extract more oil. The rest could be sold, providing profits to the oil companies and supplying most people with cheap fuel for the cars that started to become commonplace.

We've now used up these good deposits. Today, we have to save about one twentieth of the oil we extract from conventional oil fields (the ones that can be extracted without specialized technology) in order to maintain production. The more difficult, unconventional sources such as shale gas, oil sands, and deposits deep in the ocean require around twice as much, or produce half as much surplus.

There are sources of fossil energy that are so deep and difficult to get to that all of the energy in the oil being pumped up (plus a bit extra) has to be used to pump up more. At this point, extraction is futile. It's almost like paying to have money in the bank. This type of energy source should definitely stay in the ground forever.

OUT OF THE FOSSIL SOCIETY

As with all other geological resources, it's difficult to know exactly how much energy can still be extracted from Earth's crust. A lot of it depends on how effective the extraction technology can be, and how much we're willing to pay, both in terms of energy and money. However, most people agree that we've already used up significant amounts of our fossil energy resources, and that the age of oil will come to an end during this or possibly the next century.

In addition, there's a growing consensus that the carbon emitted into the atmosphere when we burn fossil energy sources is in the process of changing Earth's climate. Increasing temperatures, too much or too little rain, other extreme weather patterns, and acidic oceans will have serious consequences, both for us humans and for all natural ecosystems. This will also contribute to reducing nature's ability to capture solar energy and use it for maintaining a healthy environment. It may therefore be best to require even more of the fossil energy sources to remain underground. But what will we do without oil?

GEOTHERMAL HEAT AND NUCLEAR POWER: ENERGY FROM EARTH'S BEGINNINGS

Earth also has other energy sources to offer. Some of the heaviest elements in Earth's crust have atomic nuclei that break down every once in a while. These radioactive materials can form when neutron stars collide either with each other or with black holes, and now make up a small part of the material that Earth consists of. The heat produced when the atomic nuclei break down warms up Earth's crust and flows out to the surface. We observe this when we dig mines; the deeper we dig, the hotter it gets.

It's possible to utilize the heat in Earth's crust to warm up our houses or even produce electricity. In most places on Earth, however, the heat current is too small to be of much use. It's only in special places—such as volcanic areas or along the mid-ocean ridges, where Earth's plates split apart and open up a gap filled with hot, molten rock—that society could potentially count on being powered by geothermal heat. This applies to Iceland, for example, which is a major aluminum producer because of cheap electricity created by geothermal power plants.

The reason the heat flow from Earth is so small is that the radioactive materials release their energy so slowly. Over the past decades, however, humans have developed methods for making these processes go faster. The reactor in a nuclear power plant is designed so the splitting of a single nucleus of the radioactive element uranium will cause another uranium nucleus nearby to break down as well. This causes more and more reactions to take place, and the heat from these chain reactions can be captured and used to produce electricity.

The nuclear reactors being built and operated today make use of only a negligible fraction of the total amount of energy available in the raw material. Energy production based on today's nuclear technology could continue for between 60 and 140 years before we run out of uranium, and will therefore not make much of a difference to our society in the long run. At the same time, radioactive waste from the reactors can be harmful to humans and the environment for a long, long time, even if the total volumes of waste are small compared to all the other toxic waste being produced from mining and industry.

It is possible to build reactors that utilize much more of the energy in the radioactive materials we extract. New technology may make it possible to live on nuclear power as much as 25,000 years into the future. However, the alternatives being developed today place such great demands on the materials in the reactors that a safe and durable solution has yet to be found. In addition, these reactors can produce material that is perfect for making nuclear weapons. In this way, we can produce energy for our society, but at the same time we may be facilitating the complete annihilation of humanity should the power plants fall into the wrong hands.

POWER STRAIGHT FROM THE SUN

The only true long-term energy solution would be to make use of the energy that is constantly streaming to us from the sun, and to do so in a way that doesn't diminish our ability to utilize solar energy in the future. Earth receives thousands of times more energy from the sun than what we currently use for electricity, industry, and transport. All we need are methods for capturing a small portion of this energy and guiding it through our civilization.

Solar cells are devices that can convert solar energy directly into electrical energy. The solar cells we're most familiar with (for example those found on an increasing number of residential roofs) are made of silicon crystals. Sunlight that hits the solar cell causes electrons to detach from the silicon atoms, and the solar cell is designed so that the electrons can only return to their atoms by detouring through an electrical circuit. These moving electrons are electrical current, which we can use to charge batteries and operate refrigerators. The development of solar cells has seen rapid progress in recent years, and many believe this will be the most important technology in our journey out of the age of oil. The question is: Will we have enough materials to actually build all the solar cells we'll need?

Silicon is no problem; it's found in every rock on Earth. However, most solar cells today also contain lead, silver, and tin. Calculations suggest that the supply of lead from mining may diminish before the year 2050, followed by tin and silver a few decades later. Newer and potentially better types of solar cells also contain rare elements such as gallium, tellurium, indium, and selenium. These elements occur in Earth's crust along with the large metals and are produced together with them. We therefore only get selenium when we extract copper, and the price of gallium is closely linked to aluminum production. As with all our metal usage, the question of whether we have enough is a matter of priority.

It's also possible to create solar cells with dyes, which nature has designed to capture certain parts of sunlight and use solar energy to move electrons. This occurs in photosynthesis, with the help of green chlorophyll. One advantage to this solution is that the dyes can be made from living organisms, and that they consist largely of carbon, hydrogen, and oxygen—elements we have in abundance all around us. However, they still need to be used in conjunction with other substances such as titanium oxide in order to actually utilize the electrical energy that can be produced from the dyes. Dyes can also be damaged by ultraviolet radiation from the sun, just as our skin can get sunburned. Molecules in living organisms are constantly being destroyed by solar radiation and other stressors, and organisms spend a lot of energy building new molecules and getting rid of those that have been destroyed. In solar cells that aren't actually alive themselves, mechanisms must be developed that can provide the same form of protection, or the solar cells won't last very long.

Solar cells work well when the sun is shining, but as we all know, Earth spins round and round, and we often find ourselves on the shady side. Fortunately, we can also benefit from solar energy in other ways.

WATER THAT RUNS, WIND THAT BLOWS

Norway's an unusual case in that it already gets almost all its electricity from the sun. Its industrial society wasn't founded on oil, but on electricity from the water the sun lifts up into the mountains. Today, my country is full of dams, pipes, and turbines that provide us with clean and renewable electricity. Hydroelectric power is also a very effective way of utilizing energy.

However, there are limits to how many rivers can be diverted into pipes. After all, landscapes and ecosystems also need running water. In Norway and the rest of the world, most are in agreement that the age of large hydroelectric plants is over. So many of the world's water systems have already been built up that new developments won't be able to take over for more than a small proportion of the fossil energy we use today.

We can expand our use of wind power, however. There's obviously an abundance of wind here on Earth—in open landscapes, in the mountains, along the coasts, and out at sea. In recent years, wind turbines (often called windmills) have become larger and more efficient. This trend accelerated in the 1970s, when the oil crisis forced the development of alternative energy sources. Many countries can produce much of—perhaps even all—the energy they need using wind turbines alone. This could generate enough power to manage without fossil energy for all the cars, heavy transport, and industry—but it would require building wind turbines, access roads, and power lines in landscapes that are deeply cherished by the locals. Thus far, there's been quite a lot of local resistance against the development of wind turbines.

While hydropower can give back as much as a hundred times more energy than what we invest, the ratio for wind is closer to twenty. This ratio depends on how much energy is needed to build the turbine, how much energy it will require to maintain it, and how long the turbine can be used before it needs to be replaced. Today's turbines have a life expectancy of about twenty to thirty years but can also be upgraded to get up to fifteen more years of life. In comparison, the life span of coal mines and nuclear power plants is somewhere between thirty and fifty years. A society driven by wind will therefore require more replacement than the fossil fuel society, but at the same time, the work and costs would be spread out over time.

A wind turbine is made of concrete and steel. Molybdenum is added to the steel to make it extra strong and then covered with a layer of zinc so it won't rust. The electrical wires are made of aluminum and copper, while the rotor blades must be made of a strong yet light material, preferably plastic-reinforced fiberglass around a core of plastic foam or balsa wood. The rotor blades are attached to the machine housing (the central part of the turbine where movement is converted to electricity) with steel. Here, there are powerful magnets, which in today's best wind turbines are made of an alloy of iron, boron, and neodymium, one of seventeen of the elements known as a "rare-earth element."

THE RARE-EARTH ELEMENTS

In reality, the rare-earth elements aren't actually all that rare in Earth's crust, but since there are so few geological processes developed to collect them in an effective way, they're difficult to extract. In neodymium and its cousins samarium, gadolinium, dysprosium, and praseodymium, the electrons are placed in a way that makes these elements invaluable in magnets and other electronic components. No one has been able to find other elements that work as well.

An ore with rare-earth minerals usually contains a mixture of several of these elements with almost identical chemical behavior. It therefore takes a lot of water, chemicals, and work to separate the different elements from one another. The same goes for recycling, since these elements are always used in alloys, often in tiny quantities compared to the other ingredients. For example, you can probably find neodymium in your mobile phone's speaker. This element must be separated from the approximately thirty other elements to be able to be used again for other purposes. This is demanding, but the technology for extracting and separating rare-earth elements safely and cheaply is constantly evolving.

If the world is to embark on a massive expansion of wind tur-bines, this could put a great deal of pressure on the supply of neodymium and its relatives. Today, China fully dominates the world's production of rare-earth elements. Brazil has the world's next-largest proven reserves but hasn't yet built up extraction on a large scale. At some point, it could prove quite problematic that these critically important elements can only be produced in a few countries.

Beneath a flat and lush agricultural landscape in Ulefoss, Nor-way, lies what is perhaps Europe's largest deposit of rare-earth minerals. The Fen Complex, as it's known, has a very special geo-logical history; it is one of the few places in the world where vol-canoes have spewed out lava that contains carbon. Most of the time, the carbon found in the rock is gone before the lava reaches the surface, and lava rocks therefore usually contain mostly silicon and oxygen. Carbonaceous lava can be formed when molten rock rapidly forces its way from the magma up to Earth's surface. This happened about 580 million years ago beneath what is today the south-central region of Norway called Telemark. On the way up, hot, carbonaceous liquid flowed through Earth's cracked crust, taking a host of elements with it, including the rare-earth ele-ments that have now become so valuable. Today, the Fen Complex is being surveyed to figure out if (and possibly how!) these resourc-es can be extracted. Perhaps we'll find neodymium from Telemark in wind turbines around the world sometime in the near future.

POWER ON A QUIET WINTER NIGHT

I'm in the process of installing solar cells on the roof of my house, but even when those are in place, I'll still need to rely on hydro-power plants for the majority of what we use electricity for at home. For most of the year, we only get sun while we're at work, and it's dark outside when we're baking bread and using the dishwasher.

In Norway, we get our electricity from hydropower, where we can open and close the water supplies powering the turbines to release just as much energy as we need at any time. The vast majority of other countries produce their electricity from fossil energy sources. If these countries were to replace coal and gas with sun and wind, would there be enough electricity for everyone on a quiet winter night?

If you were to get all your power from a rooftop solar panel and a wind turbine in the backyard, you'd be at the mercy of the weather. It would help if you were also connected to a plant in the neighboring town, since they might have more wind than you do at different times. If we were to pull power lines all over Europe, it might always be windy enough somewhere for everyone to get as much power as they need. Large computers could then use weather forecasts to predict what production in different parts of the grid will be at any time, and use historical data on power consumption to move electricity where it was needed.

But then we have to imagine: If the whole of Europe is dependent on the wind blowing in Spain, and the wind in Spain is so strong one evening that its power lines topple over, streets could go dark all the way in Oslo. In order to keep the system from becoming too vulnerable, there must be stores of energy that can be used if the supply gets too low. How can this be done?

Hydropower is one option that can allow us to store enormous amounts of energy. We can turn hydropower off and on whenever we want—but we can also run it backward. This just requires some large pumps and pipes, and there are already such plants in operation in several places across the world, such as the large Bath County Pumped Storage Station on the border between Virginia and West Virginia in the US. When there's a lot of wind, some of the energy from the wind turbines can be used to pump water up into the reservoir behind the dam. It can remain there until the wind stops and people come home from work. When all of the electric cars are plugged in for charging, the valve is opened and water flows back through the turbines to generate electricity. We get the best results through a combination of different systems. Energy can be stored in the form of motion in so-called flywheels—heavy wheels that rotate in a vacuum, kept suspended by magnets so energy isn't lost in friction—or as heat, by excess energy being used to heat up molten salt to several hundred degrees. In addition, we can use both batteries and hydrogen to move and store energy—for example, when we need energy to transport ourselves and our goods. With the exception of some trains and trams, the transport sector is dependent upon a system that can release energy to the engine without its being connected to an electrical main.

COBALT IN THE BATTERY

Gasoline is perfect for transport. We burn gasoline while we drive, and when we turn the motor off, the leftover energy remains in the gas tank, ready for the next trip.

My car doesn't use gasoline; it stores its energy in a battery. Like oil and gasoline, batteries also store energy by forcing atoms together that are reasonably satisfied to be linked to each other. Energy is released when these elements are allowed to react with elements they like even more. Of all the elements we have to choose from, lithium is the fondest of "giving away" electrons. This means that a great deal of energy can be exchanged in chemical reactions with lithium, and since lithium is also a light element, the best batteries we have today are based on lithium under the designation "lithium-ion batteries." Ions are what atoms are called when they don't have the same number of electrons as protons in the nucleus, and when lithium moves through the battery, it's short one electron.

Aside from its use in batteries, we're familiar with lithium as an element used in the treatment of depression and bipolar disorder. The usefulness of lithium in this part of medicine is related to how lithium can take part in a range of the biochemical reactions involved in the functioning of our nerves. Lithium is a relatively common element in Earth's crust but quite rare in solid minerals. Today, about half of the world's lithium is extracted from solid rock in Australia, and the other half from saltwater sources in Argentina and Chile. Estimated lithium resources are equivalent to over 1,200 years at today's production levels, but there's still much exploration to be done to find the best deposits and to figure out how they can be extracted most effectively.

In a charged lithium-ion battery, lithium ions are chemically linked to carbon. When I start up my electric car, I allow the lithium ions to move from carbon to cobalt. Because cobalt and lithium are extra friendly with each other, this releases energy. Cobalt is an expensive element and would ideally be replaced with something less rare, but its properties are so unique that it's unfortunately difficult to find a better alternative.

Cobalt mining also has a bad reputation. Almost half the cobalt on the market today comes from the Congo, where much of it is extracted using incredibly rudimentary techniques. About a hundred thousand workers—many of them children—excavate cobalt with shovels and hoes, without any security measures, in tunnels dozens of feet below the ground. I don't like the idea that my car may have been manufactured this way.

I can drive my car pretty far each time I charge the battery, which is why it's so incredibly heavy. One pound of batteries contains much less useful energy than one pound of gasoline. Since carbon atoms are so small and light, and linked together with strong bonds, a lot of energy can be released from one pound of carbon molecules. Since the energy in gasoline is released when carbon reacts with oxygen (which is obviously everywhere in the air around us), there's no need to use up any weight or space on putting oxygen into the gas tank. In my car, I need both carbon and cobalt. If scientists managed to make batteries in which lithium reacts with oxygen in the air instead of solid cobalt, they could potentially get a pound of battery to contain almost as much energy as a pound of gasoline. Then, electric aircraft and trucks could have just as much load capacity and range as today's fossil solutions—without needing cobalt.

In other words, even if we're able to overcome the greatest technological barriers, we'd only be able to achieve something almost as good as what we have today. This is a bit disappointing. Shouldn't the future offer us flying cars and charter trips to other planets? To achieve this, we'll need energy carriers that are even more energy-dense than batteries.

Hydrogen could be such an alternative. One pound (0.5 kg) of hydrogen gas contains almost three times as much energy as one pound of oil. If you mix hydrogen gas with oxygen gas, a tiny spark is enough to ignite the mixture, releasing a tremendous amount of energy in the form of heat, while oxygen and hydrogen turn into water. On particularly sunny days, some of the electricity generated in solar power plants can be used to separate water molecules, and the resulting hydrogen gas can be stored for later use. The problem is that hydrogen gas takes up a lot of space. Storing a pound of hydrogen gas requires a balloon that can hold nearly 50 gallons (180 L), while a pound of gasoline takes only 0.17 gallons (0.5 L). In today's hydrogen cars, a lot of energy is used to condense the gas into a smaller volume, and space shuttles use liquid hydrogen cooled to $-423°F$ ($-253°C$). It takes an extreme amount of energy to get something to stay so cold, so this isn't really a feasible solution for passenger cars.

Hydrogen cars don't burn hydrogen in their motors but convert chemical energy directly into electrical current in a fuel cell. Most of today's fuel cells contain platinum, which helps separate hydrogen molecules and release electrons. Platinum is one of the rarest metals in Earth's crust and is extracted primarily as a by-product of copper and nickel mines. South Africa is by far the world's largest producer and has the largest platinum resources. In 2017, there were only four other countries with any notable platinum production. Because production is so dominated by only one country (which incidentally is also plagued by frequent mining strikes and other political issues), platinum is among the elements that authorities in several countries keep an extra close eye on. Hydrogen is therefore not a solution to all problems, but it will almost certainly be a part of the solution to our energy challenges.

GASOLINE FROM PLANTS

Growing plants capture solar energy, and this energy can be released in our engines and boilers for the benefit of society. Bioenergy is being produced as one of the new renewable energy sources that could help us enter the new age. How far can this solution take us?

In a way, bioenergy is old news; it's been humankind's most important source of energy through most of our history. The use of bioenergy for metal production and other industry has led to massive deforestation in several parts of the world. This happened at a time when Earth's population was less than a tenth of what it is today, and when each individual human used far less energy than you or I.

Waste from forestry and agriculture may seem like an alluringly cheap energy resource, but this waste is also an asset for natural ecosystems. Organic material functions as storage for carbon, bioavailable nitrogen, and other nutrients. It also helps break down harmful substances in the environment, provides shelter for small creatures on the forest floor, suppresses floods, hinders soil erosion, and provides us with cleaner air, water, and soil. If we remove too much organic material from soils and forests, we'll have to use energy to supply them with fertilizer and replace the services we get free of charge from ecosystems today.

Logging debris can't simply be put directly into a fuel tank, either. It takes a lot of energy to break down the large molecules in wood to turn them into an energy-dense liquid form that can be used as fuel. In fossil energy sources, nature has used high pressure, high temperature, and millions of years to do the same.

It's easier to produce liquid fuel from plants that naturally contain a lot of oil, such as soybean oil or palm oil, or a lot of sugar, such as sugarcane or sugar beets. But do we actually get more energy from the finished fuel than what we used to make it? The tractor that plows the ground and harvests the crops needs fuel. Seed and fertilizer production also requires energy. Seeds, fertilizer, and water need to be transported to the fields and the crops transported away. After the plants are harvested, they need to be dried, ground, heated, centrifuged, and distilled.

When energy-rich plants are grown in areas with a lot of sun, it's possible to regain as much as fifty times as much energy as was invested. For most of the biofuels on the market today, however, you can only get somewhere between two and five times as much energy. For more difficult resources such as wood, you get back about as much as you put in. In that case, biofuel production merely becomes a method of turning "black" fossil energy into "green" more than it's a method of extracting energy from nature.

In the future, we may be able to produce biofuels by growing algae in pipes or tanks in sunny areas, although thus far no one has been able to make this work on a large scale. The efficiency of such systems is ultimately limited by photosynthesis itself, which is designed so that no more than 12 percent of the sunlight that reaches the plants can be stored as energy. Solar cells of the same area could convert about 20 percent (or even more) of the same sunlight directly into electrical energy.

TODAY WE EAT OIL

Food used to be a source of energy, but today food production is actually a *waste* of energy. If you read the nutritional labels of the food you have on the table, you can see how much energy they contain. Then you can think about the fact that there might have been as much as ten times as much energy used—mostly in the form of fossil oil, coal, and gas—to produce this food. Energy is used to build infrastructure; produce and transport fertilizer, pesticides, and seeds; plow fields; operate irrigation systems; dry crops; and transport raw materials—not to mention for the production, packaging, refrigerating, transport, and preparation of the finished food.

For a long time, we've used almost all available areas for food production. Only a few decades ago in my own country, for example, every patch of land in the Norwegian fjord landscape was used for growing livestock feed, in addition to pasture grazing. Today, the steepest and most inaccessible landscape is fallow. Since the 1950s, world food production has increased dramatically, not because more land is being cultivated, but because methods have been developed that supply much more energy (for example in the form of fertilizers) for use in agricultural efforts. Today, the world's food production banks on an abundance of cheap fossil energy. In the future, we'll need even more energy to produce the same amount of food we do today since erosion, climate change, topsoil depletion, and the loss of groundwater resources will make food production conditions increasingly difficult.

This is where we find ourselves today. The fossil energy sources are being depleted. We know we have to stop burning coal, oil, and gas if we want to ensure a livable climate for our descendants, but we're still nowhere near being able to replace all the coal power plants with renewable sources for electricity, not to mention getting enough energy for industry, transport, and food production. A society without an abundance of energy cannot sustain complex structures, advanced industry, and the research we need to find solutions to the challenges of the future. Something needs to be done—and fast.

9 | Plan B

AM I TOO CYNICAL? It seems like I see problems at every turn. Too much garbage. Too little food. Too little energy. Too expensive steel. Am I simply stuck in today's way of thinking? Haven't we humans repeatedly shown we can attain the unattainable?

Among all the great visions of the future I haven't yet mentioned, there are three that deserve some attention here: first, the possibility of infinite, cheap energy; second, the ability to extract resources in outer space; and finally, the ultimate plan B: leaving Earth and starting anew on other planets.

UNLIMITED ENERGY: A SUN ON EARTH

The energy that is constantly streaming toward us from the sun is released within the sun's interior through fusion reactions. This is energy left over when the nuclei of the lightest elements (such as hydrogen and helium) merge and become other, heavier elements.

If we had a machine that could make the same thing here on Earth—a fusion reactor—we'd be able to produce huge amounts of energy from the elements we have in abundance.

However, this is easier said than done. In order for nuclei to fuse, they must be pressed together with tremendous force. The inside of the sun is about 27 million degrees Fahrenheit (15 million degrees Celsius) and has a pressure equivalent to 340 billion times the air pressure on Earth's surface. These conditions are beyond those that could possibly be recreated in a reactor here on Earth.

The task becomes more surmountable if ordinary hydrogen, which only has one proton in its atomic nucleus, is replaced by heavier versions of hydrogen that have one or two additional neutrons. These variants are called deuterium and tritium. Deuterium atoms weigh twice as much as ordinary hydrogen, and when it takes hydrogen's place in water molecules, we get what is called heavy water. This form of water was produced by Norsk Hydro and was the reason for sabotage actions against the plant during World War II; heavy water is useful for making nuclear weapons from plutonium. (It's worth watching the 1965 film *The Heroes of Telemark*, if you're not familiar with the story.) Tritium, the two-neutron hydrogen variant, is an exceedingly unstable substance that breaks down into other elements within a few years of its formation. If we wanted to use tritium in fusion reactors, we would first have to make it ourselves. Today, tritium is produced by a rare variant of lithium that makes up less than 10 percent of all lithium on Earth.

Hydrogen may be an endless resource, but lithium isn't. Calculations indicate that if we were to use all the lithium that could possibly be excavated from Earth's crust in fusion reactors, it would be able to sustain today's energy consumption for about a thousand years. In addition, both deuterium and lithium can be found in seawater. With an effective method of extracting these elements from the sea, we'd have enough of them to cover human energy needs for millions of years to come.

In a fusion reactor, the electrons must be detached from the atoms such that the atomic nuclei can get close enough to each other to fuse. A gas so hot that the electrons have detached from their atoms is called a plasma. Here on Earth, we find plasma in lightning and the Northern Lights, among other things. The problem with plasma is that it tends to spread out and, therefore, cools down quickly. Stars are so large and heavy that their gravitational field can hold the hot plasma in place—but this can't be recreated here on our little pebble of a planet. Our alternative is to use magnets to capture the plasma in a magnetic field of a particular shape. If the reactor is designed in such a way that the plasma never comes in contact with the walls, the heat in the plasma won't be lost to its surroundings, and the reactor walls won't melt or burn up.

The pursuit of fusion energy started during the Cold War on both sides of the Iron Curtain. In 1968, scientists from the Soviet Union reported that they'd produced high-temperature plasma in a donut-shaped magnetic field called a tokamak. Shortly thereafter, British scientists had achieved the same. Today, scientists from around the world are collaborating to build the world's largest fusion experiment, ITER, in France. If all goes according to plan, ITER's tokamak will produce its first plasma in 2025.

The problem with the tokamak is that it has to be controlled incredibly precisely, and the magnetic field is maintained only if the electrical current in the system increases and increases . . . and increases. This obviously can't keep up for long, and the engineers at ITER expect to be able to hold on to the plasma for about half an hour before the machine has to be shut down and cooled off. These constant temperature fluctuations will be incredibly demanding for the materials to be used in the reactor.

In an alternative design with the futuristic name "stellarator," the magnetic field has an extremely complicated shape that allows the machine to run without interruption. Such a reactor was first proposed in the 1950s, but it wasn't until the 1980s that computers became powerful enough to allow physicists to get down to the business of designing its complex geometry. The stellarator Wendelstein 7-X in Germany met the goal of keeping hydrogen plasma at over ten million degrees Celsius for about one second in 2016, and engineers are now working on upgrading the reactor.

Nuclear fusion will only be possible as long as the high temperature in the reactor is maintained. If something were to go wrong and the reactor lost control over the magnetic field, everything would stop. There is therefore no danger of uncontrolled reactions, explosions, or meltdowns like the ones we know from the accidents at the nuclear power plants in Chernobyl and Fukushima. Fusion reactors would also create far less waste than today's nuclear plants, but when the neutrons emitted through fusion hit the materials in the reactor, some radioactive waste will form that must be treated as special waste for several hundreds of years.

Although development is slow, there isn't really any reason why we shouldn't succeed in building fusion power plants. Seawater may very well be our most important energy resource in a hundred, two hundred, or five hundred years. All that's required is enough money and resources to run major research programs until we succeed. But the question remains: Would an almost inexhaustible source of clean energy really mean the end to all our problems?

Not entirely. Energy is indeed our number one demand, but our other material needs aren't automatically covered by having enough energy. Bike gears will still wear down and steel pylons will still rust. To replace what we lose, we'll need to keep crushing ever-increasing volumes of stone. The holes in Earth's crust and the piles of waste don't just become new rock again, no matter how much energy we have available. Perhaps we would have enough materials in a world that's been completely dug up and pulverized—but that's not exactly something we want.

What's more is that we'll still need food, air, and water. Clean air, clean water, and fertile soil require more than just energy; they demand ecosystems that work, and ecosystems depend on sufficient amounts of the right type of nutrition and the ability of large chains of vulnerable mechanisms to work together.

ELEMENTS IN SPACE

We may very well reach a limit to how much of Earth's resources we should extract. But then again, why should we only limit ourselves to Earth? We do have a whole universe to take advantage of.

We've already been using materials from space—just look at Tutankhamen's dagger made of iron that had fallen from the sky. Tens of thousands of meteorites hit Earth's surface each year, although most of them are the size of a speck of dust. We get around 2,500 metric tons of iron, 600 metric tons of nickel, and 100 metric tons of cobalt from outer space each year. In comparison, we extract 1.5 billion, 2 million, and 110,000 metric tons of these metals from Earth's crust every year, respectively.

It is true that outer space is full of materials. In our solar system alone there are thousands of asteroids—objects that orbit around the sun but are much smaller than the planets. The smallest are about the size of a pebble, while the largest known asteroid, Ceres, has a diameter of approximately six hundred miles. Most asteroids are found in a belt between Mars and Jupiter somewhere between 200 and 300 million miles (300 and 500 million km) from Earth. There are most likely also several thousand asteroids that are closer to the Earth; 250 of them are known today, but it's been estimated that over 1,000 significantly large objects will get close enough to Earth to be able to hit it sometime in the future.

Asteroids are far away and difficult to study. What research-ers know about asteroid construction has been learned by looking at the light reflected from their surfaces and sent here to Earth, or from images taken by space probes that have passed relatively close to asteroids. They can also study the meteorites that have actually landed here on Earth and presume they bear some resem-blance to asteroids.

So far, scientists know that about three quarters of known aster-oids consist of carbon, oxygen, and other elements that are com-mon on Earth. They also contain considerable amounts of water in the form of ice. The next most common type of asteroid consists largely of iron, silicon, and magnesium, while just under 10 per-cent of asteroids contain metallic iron along with other valuable metals such as cobalt, gold, platinum, and palladium.

The total amount of raw materials in the solar system is im-mense, and there are already commercial companies with plans to start mining them. It will be easier to mine materials from an asteroid than from a planet or a moon since asteroids are smaller and, therefore, have almost no gravitational field. This means that spaceships don't need to use energy in order to brake when land-ing on the surface, and perhaps even more importantly, they don't have to use loads of energy to lift themselves and any extracted materials up from the surface and out into space. Mining from an asteroid would involve breaking out the materials, sorting out the desired minerals in some kind of rotating wheel, and then allow-ing the material to float up on its own or be collected in a gigantic net to be towed back to Earth.

The main disadvantage is that asteroids are so far away from us. If people are going to be sent off into space as miners, they need to assume they'll be gone for at least a few years. Only a few people have spent around a year in space, and the results aren't encouraging. Long periods of weightlessness impair muscle, blood, balance, and vision. The alternative to sending astronauts could be using unmanned spacecraft to do the whole job. An easier option could be to get hold of a whole asteroid and tow it somewhere closer to Earth—for example, into orbit around the moon—and then send astronauts up on shorter trips to dismantle it.

Perhaps in the future we'll actually get everything we need from outer space rather than from mines here on Earth, thus freeing our ecosystems from further burden. We could use fusion energy to produce the fuel needed to send spacecraft out into space and bring back raw materials. This is a good idea in theory, but a tremendous challenge in reality. Mines here on Earth can be operated by almost anyone who has a pickax, a spade, some chemicals, and access to fuel. Space travel is an enormously costly and complex operation reserved only for the very rich. A society in which we got everything we needed from outer space would have to be organized quite differently from the world as we know it today.

We still aren't entirely ready to transport resources from asteroids back to Earth, but it has been done once before: In 2010, a capsule arrived on Earth containing a few specks of dust from the asteroid Itokawa that had been collected by the Japanese vessel Hayabusa. In June 2018, news broke that its big brother Hayabusa2 had managed to land on the asteroid Ryugu. Now, scientists in the Japanese and American space programs hope they'll be able to get a few grams of material from Ryugu in 2020. NASA sent off its own spaceship, OSIRIS-REx, in 2016. It reached the asteroid Bennu in December 2018, and it's planning to use a robotic arm to dig up dust and stone from the surface and collect several pounds of materials, for expected arrival back on Earth in 2023. Each of these expeditions to the asteroids closest to Earth takes about seven years to collect nothing more than a few pounds of unspecified materials.

A future based on elements from outer space will, just as with the development of fusion energy, require the maintenance of large, expensive research programs over a long period of time. In other words, it's far from a simple solution to problems that can arise in the near future.

The commercial companies eager to make money in space have also realized that it could be a long time before they can earn a profit selling gold and cobalt from asteroids and have instead shifted their focus to resources that could be useful in space. It's expensive to send materials up from Earth's surface, and if astronauts are going to be in space for a longer period of time, it will be worthwhile to be able to mine the water and oxygen they need outside Earth's atmosphere. Companies can mine ice from the most common asteroids and place it in depots in strategic places in the solar system where it can be used by astronauts on future voyages. With the help of solar energy, the ice can be used to produce both oxygen and hydrogen, which can be used as fuel for spaceships. In the short term, resources from space will therefore stay out there and not be used to build the infrastructure down here on Earth.

AWAY FROM EARTH?

If we're to believe some of the most visionary thinkers today—such as Elon Musk and the late Stephen Hawking—the future of humanity does not lie here on Earth. It's high time that we humans move on to other planets, since it looks like the planet we started on won't be able to take care of us for many more generations. If you're of a more pessimistic disposition, we have more than enough problems: Climate change, the collapse of ecosystems, and source depletion could, all on their own, be enough of a reason to move out.

People moving into space to escape a dying planet has been the subject of a number of films and books. In August 2016, I was at the library at the University of Oslo to watch a lecture with physicist Kip Thorne, who went on to win the Nobel Prize for his work on gravitational waves the following year. The theme of his 2016 lecture was the physics behind the movie *Interstellar*. In this film, a group of brave astronauts are sent out into space in search of a new planet where humans can live after all of the ecological disasters that had ravaged Earth.

The biggest challenge for scientists in the movie wasn't finding a suitable replacement planet or discovering a wormhole in space that allowed the astronauts to travel there and back without needing thousands of years. The hardest thing was getting lots of people off the ground. Our planet's gravity is formidable. Sending just a few tons of satellites into space requires rockets with enormous fuel tanks. There simply wouldn't be enough energy for everyone to make the trip.

In *Interstellar*, this task was made easier by the fact that most of the people on Earth had already died of hunger and misery. Then, at the end of the movie, our hero meets some multidimensional beings in the center of a black hole who help him understand how to turn off Earth's gravity for just a moment, thus allowing huge colonies of humans to be sent into space.

"It's a wild idea, but we don't know for sure it's impossible," Kip Thorne told the audience at the library. This is true: It's impossible to prove that something is impossible. At the same time, however, I don't think this is something that we can reasonably stake our futures on.

There are no other habitable planets in our solar system. Life spent hundreds of millions of years transforming the planet we inhabit to one where we can live—with the earth, water, just enough oxygen that we can breathe, and an ozone layer in the outermost atmosphere that protects us from hazardous radiation. Our muscles, skeletons, and blood vessels are precisely adapted to the gravity we have here on Earth. We can build colonies deep below the ground on Mars and hope that we can learn to change the atmosphere of an entire planet so that it could someday be habitable for plants and animals on the surface. However, considering how poorly we understand how to take care of the planet we already have, this proposition seems improbable.

Perhaps there are already habitable planets in the universe orbiting around other stars. The closest neighbor star is four light years away—a hundred million times further than the distance to the moon. It would take hundreds of years to get there, so the people who would have any chance of arriving there would be the distant descendants of the astronauts we sent away. They wouldn't have made the choice to travel anywhere. They also wouldn't know much about the planet they arrived on, given that we know so little about the planets that are so many light-years away. And perhaps most importantly, they wouldn't even be able to call home to ask for help when they finally arrived, since it would take so many years for the signal to reach us here on Earth. Is this the future we want to give to those who come after us?

All in all, no matter how big our problems here on Earth may seem, these alternative plans seem much, much more difficult. And even if we did ever get the chance to develop technology for clean energy, resources from space, or settlements on other planets, we still have to first manage to maintain a rich and complex society here on Earth for many decades or centuries to come.

10 | Can We Use Up the Earth?

WE CAN'T USE UP EARTH. In fact, to "use up" might be the wrong expression. Everything we have on Earth remains here on Earth (apart from the very, very lightest gases). Helium disappears into space when we use it (a concern in itself, and perhaps a good reason not to buy those helium balloons for your next birthday party), but all of the iron, aluminum, gold, and carbon we use in our things and to build our bodies enter into different circuits and stay on our planet for all time. Earth's circuit will make sure to clean, collect, and ready everything for use in the next round—for those who have the time to wait.

And time is something we don't always have. Water, yes. Air, most likely. Plant materials are also a renewable resource as long as we can sow new seeds in fertile soil. But iron and aluminum? Not a chance. It will take millions of years for nature to reassemble all the iron atoms we've spread out across Earth and in the seas and to form deposits so we can dig out the iron once again. This requires volcanoes and shifting continental plates. Most of the iron we extract comes from a crucial change in the history of the planet, when life started producing oxygen for the first time, and this isn't something we can expect to happen again.

When these cycles become too slow, the resources in practice become nonrenewable for us. What's lost is lost, and what is brought forth from the crust can't be mined again.

LIMITS FOR GROWTH

In 1972, the book *The Limits to Growth* put the spotlight on just this question. Among the authors was Norwegian physicist Jørgen Randers, who later became an economics professor and president of BI Norwegian Business School. The book presented the results of a computer model for the entire world economy that was used to calculate various scenarios for how commodity prices, food production, pollution, and population growth would develop in the future—assuming that resource extraction will require more and more energy and money as the best deposits are emptied.

The most salient conclusion of the book was that continued growth will inevitably cause us to encounter a number of natural boundaries, both for how much we can extract from different resources and for how much of our waste the planet is capable of handling. If population and our standard of living continue to increase, the model indicates that society would be forced to devote an ever-increasing share of its resources to sustaining growth. At some point, perhaps even sometime during the twenty-first century, society would no longer be able to maintain the necessary level of production. Decrease in production would lead to economic decline, reduced food production, poorer living standards, and, in the long term, a reduction in population.

The authors of *The Limits to Growth* set an optimistic tone. Their models showed that if society decided to slow down and gradually stop population growth (preferably by making it attractive for people to have fewer children, for example through education, care for the elderly, and financial security) and production growth (well before we step too far beyond the natural boundaries), then this could occur in a gentle fashion, without a major burden to most people. On the other hand, if we chose to ignore the fact that these boundaries exist until it's clear they've already been passed, it could lead to a much more dramatic outcome, with economic collapse, famine, and uncontrolled population decline. In 1972, we still had plenty of time ahead of us, and the world's common sense and wisdom might have been able to successfully complete the project in a reasonable way.

Millions of copies of the book were sold around the world, sparking a widespread debate. Critics believed—and still do—that the book is based on a simplified portrait of reality. The world economy doesn't follow such simple laws as the ones the data model was based upon. The book neglected the ultimate resource, critics say: human intelligence. If there's a scarcity of a natural resource, we find new and better recovery methods. If we have too little oil, we get the energy we need from solar panels and wind turbines. This is how we can always work ourselves out of a difficult situation. For us, no boundaries are truly absolute.

The year 1972 wasn't the first time academics warned the world of the impending scarcity of resources and imminent societal collapse. Throughout history, the concern over limits has been brought up time and time again. Among the most well-known pessimists is Thomas Malthus, who in 1798 cautioned that population growth would lead to a famine. In 1766, Adam Smith, known as the founder of the subject of social economics, described how man, like all other creatures, must remain within certain natural boundaries. Perhaps this long history of painting such a gloomy picture is itself an argument to put an end to these doomsday prophecies once and for all; to be fair, it has gone well so far. We humans have already demonstrated that we don't need to concern ourselves with natural limitations.

Can this be true? Is it right to think that since everything has gone well up to this point, it will continue to go well forever?

GROWTH THAT IS HAPPENING FASTER AND FASTER AND FASTER

Growth in society, be it population growth or economic growth, is usually described as a percentage. That the economy is "growing by 2 percent a year" describes the same thing that happens when you get interest on the money you put in the bank. For example, if you put $100 into a good savings account where you get a 10 percent interest rate, after one year you'll have $110 in the same account. The profit that year was 10 percent of $100, which is $10. A year later, you'd have a profit of 10 percent of $110, which is $11. Then the account balance has increased to $121. The year after, you'll earn $12.10—and so on. Although the interest rate is the same, you make more and more money each year (assuming you don't take any money out of the account). After eight years, you'll have more than $200 in the account, because a 10 percent interest rate causes the initial value to double after seven years and four months, no matter how much money you start with in the account.

The type of growth in which something increases by a certain percentage each year is called exponential growth. When something grows exponentially, the total will always increase faster and faster and faster. Although growth has varied over time, this is the situation both for the number of people on Earth and for how much of Earth's resources we use.

Imagine you're sitting in a large sports arena in one of the nose-bleed seats.

All the doors in and out of the stadium are closed.

Then, someone places a special drop of water in the middle of the field. Magic causes this drop of water to double in size every minute.

In the beginning, not much happens. It takes about twelve minutes for the drop to become large enough to fill a water glass. You can't even see it from the stands yet. After forty-four minutes, the stadium is only half filled, and you think you still have plenty of time to get away before you'll even get your feet wet.

But since there is twice as much water every minute, the stadium is completely filled just one minute later. Five minutes earlier, the water still took up only 4 percent of the volume of the stadium, and at that point, you didn't realize anything would happen.

In a system of exponential growth, we're always in a very special time. Although growth is steadily occurring at the same percentage each year, the absolute growth is greater and greater. Every year, we have more than ever before. Although everything has gone well so far, it doesn't mean it will continue that way in the future. One day we'll be sitting with our heads underwater and the sense that this happened all of a sudden.

Perhaps we are the ones who are living in the age when we're quite seriously hitting our heads against the planet's limits. At the very least, our close descendants will be. It doesn't mean that those who warned us long ago were wrong; it just means that if we'd listened, we would've had better conditions today.

THE NECESSITY OF ECONOMIC GROWTH

As a physicist, it seems obvious to me that there should be natural limitations. It's therefore been difficult for me to understand why economists keep insisting on continued growth. I agree that the poorest countries in the world are in dire need of allowing their citizens and government to increase consumption in order to improve the standard of living. In Norway, we have one of the highest standards of living in the world. Don't we have *enough* already? Isn't it time to work less and have more free time? Why should our economy continue to grow?

The answer turns out to be in the economy itself. Our economy is built on growth, and many would say that the alternative to growth isn't a plateau, but a collapse.

At some point in time, money was a measure of absolute value. A gold coin was worth something as a coin, but also as gold. Later, gold coins were replaced by coins made of other, less valuable metals and paper notes, which essentially have no real value in themselves. These were symbols of a certain amount of value—preferably gold—that lay in a bank vault. Instead of people carrying around precious gold, they could trade with money as a symbol of gold.

This isn't the case today.

Imagine that you want to start a bakery in your neighborhood. It costs money to enter into a lease agreement, buy equipment, hire a chef and waiters, and market your new business. You do, however, expect that after a certain amount of time, you'll get more money from the bakery operation than you need to cover all of the expenses it takes to run it. You therefore go to the bank to ask for a loan.

The bank reviews your plans and agrees to lend you what you need to start the business. This isn't something the bank does out of the goodness of its heart; it's a business that sees an opportunity to increase its value. Had it kept the money to itself, it would have retained its original value, but if it lends money to you, you'll use it to start up a business ideally worth even more than the original value of the money. After a few years, you can pay back what you borrowed from the bank as well as a good deal of interest. The bank makes money on the loan, you earn money on the bakery, and everyone's happy.

This is the principle that drives our economy forward: We trust that things will go better in the future. The value of a banknote isn't in gold but in hope for the future. When we're young, we don't wait to buy an apartment until we've saved up enough money to pay cash; we take out loans from the bank that will end up costing us a lot more than the actual purchase price but that still make sense because we expect to earn higher wages in the future. Companies get the money to replace old equipment and start new operations by selling shares, which are attractive to buyers as long as they expect the company to succeed in the future, thereby increasing the value of their shares. Where the alternative might have been for everyone to have a bag of gold for their own use, our economy is structured in such a way that it pays to let someone else spend your money when you aren't. This is how society's resources are used where they can create products and services for the benefit of the people. The economy grows when your bakery sells more cinnamon rolls each month, thus allowing you to earn money yourself, replace broken coffee cups, and also pay interest to the bank.

Now, let's say that the economy isn't growing, but shrinking. You're selling fewer and fewer cinnamon rolls a month. Then it gets hard for you to make enough money to replace all the broken cups or the worn-out oven. If you ask the bank to help you out with a new, small loan, they'll take a look at your accounts and figure out that you'll have problems paying interest, and you'll get rejected. This is, of course, bad for your business. You might have to shut down and lay off the bakers and waiters. This leads to their having less money to buy cinnamon rolls at the neighboring bakery, which, in turn, also has problems keeping its operations going—and things start to snowball. Economic downturns lead to unemployment and social unrest. For those governing the country, it's important to avoid such conditions.

According to some theories, a constant economy can't be maintained over time, either. If you want a loan for a new oven, it's not enough to simply show you're selling the same number of cinnamon rolls each month. The bank will say that if you couldn't afford the new oven today, you won't be able to in a year or two, either. Without expected growth, it makes more sense to loan money to or buy shares in the neighboring bakery, which has just opened a new branch with hungry customers. The world economy only goes up or down, and not straight ahead, which is why even those of us living in the wealthy Western nations have to work for economic growth.

CAN THE ECONOMY GROW WITHOUT USING MORE RESOURCES?

Economic growth can be an absolute condition for an international society with social stability and prosperity. At the same time, even the most enthusiastic resource optimist agrees that the total use of resources can't continue to grow forever and ever. Let's take energy as an example: The consumption of fossil energy resources can't increase much more in the future and therefore ought to decline sharply over the next few decades. This energy can to some extent be replaced by renewable energy, but if we look far ahead, there are limits to the extent of areas we can cover with solar cells and wind turbines. Continued exponential growth will eventually require more energy than both the sun and Earth's crust can supply us with. In the long run, eternal, exponential economic growth cannot be accompanied by corresponding growth in energy use.

No problem at all, optimists will respond. We can certainly have economic growth without increasing our energy use. We'll just need to use less energy for each unit of economic benefit. This is already happening all the time. For example, we've gone from using incandescent lamps, where most of the power we send through light bulbs is converted to heat, to using LED bulbs, where almost all power becomes light. Since it's light and not heat that we want, we're getting the same economic benefit (light) by using much less electrical energy. As long as we continue to make such efficiency improvements, we can grow the economy without needing more energy.

Such a disconnect between economic growth and resource growth seems incredibly appealing. We can have our cake and eat it, too: a society in which new generations of adults will always have things a bit better than their parents did, while at the same time taking care of the planet for their descendants. Problem solved!

But is it really that simple?

Suppose we've already reached the limit for energy consumption. The economy will continue to grow, but we won't use a single unit more of energy. If growth is slow—perhaps just 1 percent a year—the size of the economy will double every 70 years. At the same time, energy is constant. That is to say that 70 years from now, we'll be so much more efficient that we'll only use half as much energy on making every cinnamon roll, every haircut, every flu vaccine, and every foot of freeway. In 140 years, we'll use a quarter, and in 210 years an eighth, in 280 years a sixteenth, and in 700 years, our distant descendants will only use a thousandth of the energy we use today to wash their hair, cure diseases, and produce iron from Earth's crust.

LED bulbs are one thing, but this assumption is bordering on the absurd. Physical, chemical, and biological processes require energy, plain and simple. Even the internet uses energy; for every search you make on Google and every post you "like" on Facebook, energy-intensive calculations are being made on a computer somewhere in the world. Today, the internet requires more than 3 percent of the world's electrical energy, but in five years the proportion may have increased to 20 percent.

AN IMPOSSIBLE PARADOX?

Our resource use simply cannot increase—otherwise we're on our way toward ecological collapse, social instability, war, and suffering.

At the same time, the economy has to grow, or we're facing economic collapse, social instability, war, and suffering. Even just the act of talking about growth reduction is dangerous because it can scare people away from making investments and lead to a recession all by itself.

Mankind's future is dependent on both growth and a lack of growth, but we can't have both at the same time.

This seems like an impossible paradox. Is mankind simply doomed to fail? Is the best we can do simply to keep the locomotive going, eat cake, and look the other way until we hit the rock wall?

I can't go along with that. We can't change how physics or biology works, but the economy is a human construct. We've made the rules and we can change them. Our economic system has given us better lives than many of our ancestors had, but now we have to find something even better if we want good lives for our descendants—both close and distant.

THE HABITABLE ZONE

Hope does exist. All over the world, economics students are revolting and demanding to learn about alternative economic models that are adapted to our era and our future. Established, prominent economists have begun to question the absolute necessity of growth. In the second edition of the book *Prosperity Without Growth*, published in 2017, economics professor Tim Jackson points out that the models that predict economic collapse in the absence of growth don't take reality into account. By taking control of the economy properly, perhaps society can counteract many of the mechanisms that lead to distress and decline in a stagnant economy. When economists allow themselves to explore the consequences of zero growth, they will hopefully figure out what strategies can work to give people a good life without destroying Earth.

In the 2017 book *Doughnut Economics,* British economist Kate Raworth describes our economy as a golden doughnut. The economy has to be large enough to provide enough food, clean water, health care, education, work, and social security to all of Earth's inhabitants. It has to stay out of the hole in the doughnut, but it can't get too big, either. It must be small enough to sustain the level of pollution, the pressure on the ecosystem, and the use of resources within the planet's limits. These boundaries form the outer edge of the doughnut. Our lives are best lived in the zone between these outer borders.

In every solar system, there is a zone with possibilities for life. Not too close to the sun, where all the water boils away. Not too far away, either, where everything freezes into ice. But in between, where it's just right—that's where we have the opportunity to flourish. A golden doughnut. That's where we live, on our planet.

Acknowledgments

Thank you to the Norwegian Non-Fiction Writers and Translators Association for the grant, and to the Department of Physics at the University of Oslo, which allowed me to take a leave of absence to work on this book. Thank you to Henrik Svensen for good advice, to Anders Malthe-Sørenssen for support, and to engaged and inspiring research colleagues in Norway and abroad.

Thank you to Jessica Lönn-Stensrud, Eivind Torgersen, Åsmund Eikenes, and Frida Røyne for their thorough and critical readings of the manuscript at various stages. Thank you to Ole Swang, who checked the chemistry in the book. Any remaining errors are completely my own.

Thank you to my editor Guro Solberg and the rest of the gang at Kagge Forlag for their enthusiastic support and invaluable feedback.

Thanks also to my translator, Olivia Lasky, for this wonderful English-language edition.

Last but not least, thank you to my good friends (you know who you are) and to my fantastic family. I'm so lucky to have you all.

References

Research articles can often be difficult to read for those outside the field. So while I've drawn on them for my writing, I've often chosen to refer here to magazine articles, blog posts, or Wikipedia instead of the original articles, so that you can actually make use of this section to inspire your own further reading. In places where I've done so, I have of course made sure that the texts I'm referring to are based on solid sources. To save space, some sources that are referred to several times are cited in abbreviated form.

Recurring Sources

Arndt et al. (2017): Arndt, N. T., et al. "Future Global Mineral Resources." *Geochemical Perspectives* 6 (2017): 52–85.

Benton and Harper (2009): Benton, M. J., and D. A. T. Harper. *Introduction to Paleobiology and the Fossil Record*. Wiley-Blackwell, 2009.

Comelli et al. (2016): Comelli, D., et al. "The Meteoritic Origin of Tutankhamun's Iron Dagger Blade." *Meteoritics & Planetary Science* 51 (2016): 1301–9.

Cordell et al. (2009): Cordell, D., et al. "The Story of Phosphorous: Global Food Security and Food for Thought." *Global Environmental Change* 19 (2009): 292–305.

Courland (2011): Courland, R. *Concrete Planet.* Prometheus Books, 2011.

Gilchrist (1989). Gilchrist, J. D. *Extraction Metallurgy.* Pergamon Press, 1989.

Giselbrecht et al. (2013): Giselbrecht, S., et al. "The Chemistry of Cyborgs—Interfacing Technical Devices with Organisms." *Angewandte Chemie* 52 (2013): 13942–57.

Harari (2014): Harari, Y. N. *Sapiens: A Brief History of Humankind.* Harper, 2014.

Holmes (2010): Holmes, R. "The Dead Sea Works." mammoth (blog), February 15, 2010. m.ammoth.us/blog/2010/02/the-dead-sea-works.

Jackson (2017): Jackson, T. *Prosperity Without Growth: Foundations for the Economy of Tomorrow.* 2nd ed. Routledge, 2017.

Kenarov (2012): Kenarov, D. "Mountains of Gold." *Virginia Quarterly Review,* January 25, 2012. qronline.org/articles/mountains-gold.

Khurshid and Qureshi (1984): Khurshid, S. J., and I. H. Qureshi. "The Role of Inorganic Elements in the Human Body." *Nucleus* 21 (1984): 3–23.

Lavers and Bond (2017): Lavers, J. L., and A. L. Bond. "Exceptional and Rapid Accumulation of Anthropogenic Debris on One of the World's Most Remote and Pristine Islands." *PNAS* 114 (2017): 6052–55.

Massy (2017): Massy, J. *A Little Book About BIG Chemistry: The Story of Man-Made Polymers.* Springer, 2017.

NRC (2008): National Research Council of the National Academies. *Minerals, Critical Minerals and the U.S. Economy.* National Academies Press, 2008.

OECD (2011): *OECD. Future Prospects for Industrial Biotechnology.* OECD publishing, 2011. dx.doi.org/10.1787/9789264126633-en.

Pipkin (2005): Pipkin, B. W., et al. *Geology and the Environment.* Brooks/ Cole, 2005.

Pomarenko (2015): Pomarenko, A. G. "Early Evolutionary Stages of Soil Ecosystems." *Biology Bulletin Reviews* 5 (2015): 267–79.

Robb (2005): Robb, L. *Introduction to Ore-Forming Processes.* Blackwell Publishing, 2005.

Rasmussen (2008): Rasmussen, B., et al. "Reassessing the First Appearance of Eukaryotes and Cyanobacteria." *Nature* 455 (2008): 1101–5.

Raworth (2017): Raworth, K. *Doughnut Economics: Seven Ways to Think Like a 21st Century Economist.* Chelsea Green, 2017

Smil (2004): Smil, V. "World History and Energy." In *Encyclopedia of Energy,* edited by C. Cleveland et al., 549–61. Vol. 6. Elsevier, 2004.

Street and Alexander (1990): Street, A., and W. Alexander. *Metals in the Service of Man.* 10th ed. Penguin Books, 1990.

Sverdrup and Ragnarsdóttir (2014): Sverdrup, H., and K. V. Ragnarsdóttir. "Natural Resources in a Planetary Perspective." *Geochemical Perspectives* 3 (2014): 129–341.

USGS (2018): US Geological Survey, Mineral Commodity Summaries 2018. doi.org/10.3133/70194932.

Wilburn (2011): Wilburn, D. R. Wind Energy in the United States and Materials Required for the Land-Based Wind Turbine Industry from 2010 Through 2030. Scientific Investigations Report 2011-5036. US Geological Survey, 2011.

Young (2013): Young, G. M. "Precambrian Supercontinents, Glaciations, Atmospheric Oxygenation, Metazoan Evolution and an Impact That May Have Changed the Second Half of Earth History." *Geoscience Frontiers* 4 (2013): 247–61.

Öhrlund (2011): Öhrlund, I. Future Metal Demand from Photovoltaic Cells and Wind Turbines—Investigating the Potential Risk of Disabling a Shift to Renewable Energy Systems. Science and Technology Options Assessment (STOA), European Parliament, 2011.

1 | The History of the World and the Elements in Seven Days

Updated times for geological eras have been taken from the International Commission on Stratigraphy's "International Chronostratigraphic Chart," v2017/02, stratigraphy.org/index.php/ics-chart-timescale.

MONDAY: THE BIRTH OF THE UNIVERSE

The early history of the universe, from the big bang, and the origin of the first atomic nuclei: G. Rieke and M. Rieke, "The Start of Everything" and "Era of Nuclei," lecture notes from the University of Arizona course Astronomy 170B1, "The Physical Universe," ircamera.as.arizona.edu/ NatSci102/NatSci102/lectures/eraplanck.htm, ircamera.as.arizona.edu/ NatSci102/NatSci102/lectures/eranuclei.htm.

The origin of the elements: J. Johnson, "Origin of the Elements in the Solar System," Science Blog from the SDSS: News from the Sloan Digital Sky Surveys, January 9, 2017, blog.sdss.org/2017/01/09/ origin-of-the-elements-in-the- solar-system.

How oxygen is made: B. S. Meyer et al., "Nucleosynthesis and Chemical Evolution of Oxygen," *Reviews in Mineralogy and Geochemistry* 68, no. 1 (2008): 31–35.

The first stars and galaxies: R. B. Larson and V. Bromm, "The First Stars in the Universe," *Scientific American* 285, no. 6 (2001): 64–71.

FRIDAY: OUR SOLAR SYSTEM IS FORMED

The solar system's origin from the pressure wave of a supernova: P. Banerjee et al., "Evidence from Stable Isotopes and [10]Be for Solar System Formation Triggered by Low-Mass Supernova," *Nature Communications* 7 (2016): 13639.

"Habitable zones"—the area where the distance from the star is just right for life: NASA, "Habitable Zones of Different Stars," nasa.gov/ames/kepler/habitable-zones-of-different-stars.

The theory of the moon's origins: R. Boyle, "What Made the Moon? New Ideas Try to Rescue a Troubled Theory," *Quanta Magazine*, August 2, 2017, quantamagazine.org/what-made-the-moon-new-ideas-try-to-rescue-a-troubled-theory-20170802.

Heavy elements sank into the middle of Earth; the ones we extract from Earth's crust came later from meteorites ("Late veneer hypothesis"): Robb (2005).

The first ocean: B. Dorminey, "Earth Oceans Were Homegrown," *Science*, November 29, 2010, sciencemag.org/news/2010/11/earth-oceans-were-homegrown.

When did plate tectonics start? On various theories and results: B. Stern, "When Did Plate Tectonics Begin on Earth?," Speaking of Geoscience: The Geological Society of America's Guest Blog, March 15, 2016, speakingofgeoscience.org/2016/03/15/when-did-plate-tectonics-begin-on-earth.

SATURDAY: LIFE BEGINS

There is no precise answer for the duration and timing of "the bombarding of Earth's crust" (late heavy bombardment): Wikipedia, "Late Heavy Bombardment," updated June 19, 2018, en.wikipedia.org/wiki/Late_Heavy_Bombardment.

Earth's magnetic field arose "during the night": New results date the Earth's magnetic field to be at least 4 billion years old (before midnight), compared to the roughly 3.2 billion years that was previously estimated. S. Zielinski, "Earth's Magnetic Field Is at Least Four Billion Years Old," *Smithsonian*, July 30, 2015, smithsonianmag.com/science-nature/earths-magnetic-field-least-four-billion-years-old-180956114.

Earth's very first organisms harvested their energy from chemical compounds deep in the oceans; summary of new theories: R. Brazil, "Hydrothermal Vents and the Origin of Life," *Chemistry World*, April 16, 2017, chemistryworld.com/feature/hydrothermal-vents-and-the-origins-of-life/3007088.article.

Earliest photosynthesis; the iron in the oceans rusted and oxygen poured into the atmosphere (Great Oxygenation Event): Rasmussen (2008). As described here, photosynthesis may have started later than what I write in the book.

The composition of Earth's early atmosphere: D. Trail et al., "The Oxidation State of Hadean Magmas and Implications for Early Earth's Atmosphere," *Nature* 480 (2008): 79–83.

Global ice age up to a quarter past nine (Huronian glaciation): Young (2013).

The first life-forms on land, after the ozone layer settled in place: Pomarenko (2015).

SUNDAY: THE LIVING EARTH

First organisms with cell nuclei: Rasmussen (2008).

First multicellular organisms: S. Zhu et al., "Decimetre-Scale Multicellular Eukaryotes from the 1.56-Billion-Year-Old Gaoyuzhuang Formation in North China," *Nature Communications* 7 (2016): 11500.

New global ice age from a quarter past three, followed by complex ecosystems in the oceans ("the Cambrian explosion"): Young (2013).

First animals and then plants on land, with subsequent development of the landscape: Pomarenko (2015).

Global ice age at 6:36 (Ordovician-Silur Mass Extinction): P. M. Sheehan, "The Late Ordovician Mass Extinction," *Annual Review of Earth and Planetary Science* 29 (2001): 331–64.

Extinction at 7:28 (the end of Devon): A. E. Murphy et al., "Eutrophication by Decoupling of the Marine Biogeochemical Cycles of C, N and P: A Mechanism for the Late Devonian Mass Extinction," *Geology* 28 (2000): 427–30.

Mass extinction 8:56, Sunday night: Z.-Q. Chen and M. J. Benton, "The Timing and Pattern of Biotic Recovery Following the End-Permian Mass Extinction," *Nature Geoscience* 5 (2012): 375–83.

Mammals and dinosaurs before 9:30 and a new global warming at 9:34 (Triassic Jurassic Mass Extinction): Benton and Harper (2009).

At 11:25 (early Eocene), temperatures began to fall: R. A. Rhode, "65 Million Years of Climate Change," en.wikipedia.org/wiki/File:65_Myr_Climate_Change.png.

11:43 (the transition to the Miocene), grassy plains: B. Jacobs et al., "The Origin of Grass-Dominated Ecosystems," *Annals of the Missouri Botanical Garden* 86 (1999): 590–643.

Hominoidea from other apes at 11:45, humans from the Hominoidea, first stone tools, use of fire: Benton and Harper (2009).

Ice ages and intermediate ages since one minute and twenty seconds ago: T. O. Vorren and J. Mangerud, "Glaciations Come and Go," in *The Making of a Land: Geology of Norway,* ed. I. B. Ramberg et al., trans. R. Binns and P. Grogan, Norsk Geologisk Forening (Norwegian Geological Society), 2006.

Bonfires in daily use, Homo sapiens, Neanderthals exterminated, language, technological development: Harari (2014).

A HALF A SECOND BEFORE MIDNIGHT: THE AGE OF CIVILIZATION

The emergence of agriculture, kingdoms, written language, money, religions, scientific revolution: Harari (2014).

Copper, iron: Arndt et al. (2017).

Steel: World Steel Association AISBL, "The Steel Story," 2018, worldsteel.org/steelstory.

Humans' use of energy, including livestock, hydropower, industrial revolution: Smil (2004).

Antibiotics: R. I. Aminov, "A Brief History of the Antibiotic Era: Lessons Learned and Challenges for the Future," Frontiers in Microbiology 1 (2010): 134.

Mankind in outer space: N. T. Redd, "Yuri Gagarin: First Man in Space," Space.com, July 24, 2012, space.com/ 16159-first-man-in-space.html.

HUMANS AND THE FUTURE

The figures for the world's population throughout history are averages of the values from ten different sources summarized in Wikipedia, "World Population Estimates," updated July 21, 2018, en.wikipedia.org/ wiki/World_population_estimates. Today's population is taken from Worldometer at worldometers.info/world-population.

2 | All That Glitters Isn't Gold

HOW EARTH'S CRUST DID US A FAVOR

The geological history of Roșia Montană: I. Seghedi, "Geological Evolution of the Apuseni Mountains with Emphasis on the Neogene Magmatism—A Review," in *Gold-Silver-Telluride Deposits of the Golden Quadrilateral, South Apuseni Mts., Romania,* ed. N. J. Cook and C. L. Ciobanu, IAGOD Guidebook Series 12, International Association on the Genesis of Ore Deposits, 2004.

How gold is transported via water: Robb (2005).

THE FIRST GOLD

Gold ten thousand years ago: Sverdrup and Ragnarsdóttir (2014).

GOLD IN THE RIVER GRAVEL

Gold was humankind's first metal: Sverdrup and Ragnarsdóttir (2014).

First large-scale mining with washing pans: Gilchrist (1989).

Mining five thousand years before our era in the Carpathian region and in the Balkans: H. I Ciugudean, "Ancient Gold Mining in Transylvania: The Roșia Montană-Bucium Area," Caiete ARA 3 (2012): 101–13.

The legend of Jason and the Golden Fleece: Wikipedia, "Jason," updated September 4, 2019, en.wikipedia.org/wiki/Jason.

Use of sheepskin in gold extraction: T. Neesse, "Selective Attachment Processes in Ancient Gold Ore Beneficiation," *Minerals Engineering* 58 (2014): 52–63.

THE MINES IN ROȘIA MONTANĂ

Fire setting used by the Dacians; the Romans defeated Dacia in 106; 165 tons of gold; the Habsburgs: Roșia Montană cultural foundation: Rosia Montana Cultural Foundation, "History," rosia-montana-cultural-foundation.com/history.

Fire setting in Norwegian mines until the end of the nineteenth century: Wikipedia, "Fyrsetting," updated April 13, 2018, no.wikipedia.org/wiki/Fyrsetting#cite_note-ReferenceA-5.

The history of Alburnus Maior, four miles of Roman mines, Habsburg water-powered crushing mills, developments after 1867: Kenarov (2012).

The Romans left the area in 271: D. Popescu, "Romania and Gold: A 6000 Years Relation," Dan Popescu—Gold and Silver Analyst (blog), August 27, 2016, popescugolddotcom.wordpress.com/2016/08/27/romania-and-gold-a-6000-years-relation.

MINING ON THE SURFACE

Nearly ninety miles of mining; open-pit mining from the 1970s: Kenarov (2012). On open-pit mining and environmental consequences: Arndt et al. (2017).

A TOXIC MEMORY

How gold is separated from tailings: Gilchrist (1989).

Geamana landfill: R. Besliu, "Romania's Unsolved Communist Ecological Disaster," openDemocracy, March 19, 2015, opendemocracy.net/en/can-europe-make-it/romanias-unsolved-communist-ecological-disaster. Environmental problems, waste disposal sites: Pipkin (2005).

FROM STONE TO METAL

Use of mercury: Gilchrist (1989).

Use of cyanide: Pipkin (2005).

Cyanide in cherry pits and hydrocyanic acid: Wikipedia, "Hydrogen Cyanide," updated January 27, 2020, en.wikipedia.org/wiki/Hydrogen_cyanide.

Baia Mare accident: Wikipedia, "2000 Baia Mare Cyanide Spill," updated June 21, 2018, en.wikipedia.org/wiki/2000_Baia_Mare_cyanide_spill; United Nations Environment Programme, "Cyanide Spill at Baia Mare Romania," reliefweb.int/sites/reliefweb.int/files/resources/ 43CD1D010F030359C12568CD00635880-baiamare.pdf.

Cyanide is safe, in over 90 percent of the world's gold mines, over five hundred mines: T. I. Mudder and M. M. Botz, "Cyanide and Society: A Critical Review," European Journal of Mineral Processing and Environmental Protection 4 (2004): 62–74.

GOLD RINGS FROM A TON OF STONE

1,700 metric tons extracted, quarry mining operation ended in 2006: Kenarov (2012).

Well over 300 metric tons of gold left: Gabriel Resources, "Projects: Rosia Montana," gabrielresources.com/site/rosiamontana.aspx. The figure 300 comes from Proven and Probable Reserves: 215 Mt @ 1.46g / t Au = 314 metric tons of gold; in addition, Measured and Indicated Resources: 513 Mt @ 1.04 g / t Au = 534 metric tons of gold.

Average gold ore concentration today and 150 years ago: Sverdrup and Ragnarsdóttir (2014).

Roșia Montană is Europe's largest gold deposit: J. Desjardins, "Global Gold Mine and Deposit Rankings 2013," Visual Capitalist, February 9, 2014, visualcapitalist.com/global-gold-mine-and-deposit-rankings-2013.

THE END OF ROȘIA MONTANĂ

The battle for Roșia Montană: Kenarov (2012); Salvați Roșia Montană (Save Roșia Montană, website of the campaign against the mining project), rosiamontana.org.

Four new open-pit mines, cyanide extraction; Roșia Montană would be buried, including four churches, six graveyards: Gabriel Resources, "Management of Social Impacts: Resettlement and Relocation Action Plan," 2006, gabrielresources.com/documents/RRAP.pdf; Gabriel Resources, "The Proposed Mining Project," gabrielresources.com/ documents/Gabriel%20Resources_ProposedMiningProject.pdf.

250 million metric tons of waste: Gabriel Resources. "Sustainability: Environment." gabrielresources.com/site/environment.aspx.

GOLD AND CIVILIZATION

The price of gold, reaction to political events in 2016, what gold is used for today: US Geological Survey, Mineral Commodity Summaries 2017, doi.org/10.3133/70180197.

THE LOST GOLD

All information in this section is taken from Sverdrup and Ragnarsdóttir (2014), except the number of metric tons mined in 2016, which was taken from USGS (2018).

3 | The Iron Age Isn't Over

"The Iron Age isn't over" ("We entered the iron age about 1,500 years ago, and have never left it"): Sverdrup and Ragnarsdóttir (2014).

The use of iron led to a revolution in warfare: J. Diamond, *Guns, Germs, and Steel: The Fates of Human Societies*, W. W. Norton & Company, 1997.

THERE'S NO POINT IN BREATHING WITHOUT IRON

Iron in the body's transportation system, four grams of iron in the body: Khurshid and Qureshi (1984).

INTO THE IRON AGE

Tutankhamen's dagger: Comelli et al. (2016).

All early iron objects come from meteorite iron: A. Jambon, "Bronze Age Iron: Meteoritic or Not? A Chemical Strategy," *Journal of Archaeological Science* 88 (2017): 47–53.

Deposits of metallic iron in Greenland: K. Brooks, "Native Iron: Greenland's Natural Blast Furnace," *Geology Today* 31 (2015): 176–80.

Production of iron from iron ore: Gilchrist (1989).

Half a ton of carbon dioxide per ton of iron: Sverdrup and Ragnarsdóttir (2014).

SWEDISH IRON

Almost all of the iron ore we dig up originated 2.5 billion years ago, extracted in open-pit mining: Arndt et al. (2017).

Kiruna's history, as well as its significance for Hitler: T. Weper, "Jernmalmen i Kiruna ble det svenske gullet" (Iron ore in Kiruna became Swedish gold), *Illustrert Vitenskap Historie* 3 (2010): 50–53. The significance of iron from Kiruna for Hitler is also described on Wikipedia, "Swedish Iron-Ore Mining During World War II," updated March 31, 2018, en.wikipedia.org/wiki/Swedish_iron-ore_mining_during_World_War_II.

How the iron ore under Kiruna was formed: Robb (2005).

The relocation of Kiruna: F. Perry, "Kiruna: The Arctic City Being
 Knocked Down and Relocated Two Miles Away," *Guardian*, July 30,
 2015, theguardian.com/cities/2015/jul/30/
 kiruna-the-arctic-city-being-knocked-down-and-relocated-two-miles-away.
 The timeline for the relocation, which has already begun, can be
 found at the website for Kiruna Kommun under "Tidslinje-Kiruna
 stadsomvandling" ("Timeline-Kiruna urban transformation"), kiruna.se/
 stadsomvandling.

FROM ORE TO METAL

Ore train to Narvik today: Bane NOR, "Ofotbanen," banenor.no/
 Jernbanen/Banene/Ofotbanen.
Largest iron metal manufacturers: USGS (2018).
Production of iron: Gilchrist (1989).
The use of hammered pig iron, cast iron, and wrought iron: Street and
 Alexander (1990).
Ore from the bogs in Scandinavia: L. Skogstrand, "Det første jernet" (The
 first iron), updated October 26, 2017, Norgeshistorie.no, norgeshistorie.
 no/forromersk-jernalder/teknologi-og-okonomi/
 0405-det-forste-jernet.html; Store Norske Leksikon, "Jernvinna"
 (Bloomery), updated December 12, 2016, snl.no/jernvinna.

COVETABLE STEEL

Manufacturing of steel, costly until the nineteenth century; structure and
 properties: Street and Alexander (1990).
Vanadium, manganese, molybdenum, chromium, and nickel in steel: NRC
 (2008) and Sverdrup and Ragnarsdóttir (2014).

THE PROBLEM WITH RUST

Society spends a lot of money (5 percent of GDP, USA, 1978) on
 counteracting and repairing rust: E. McChafferty, *Introduction to
 Corrosion Science*, Springer, 2010.
How rust occurs and methods for preventing it: Street and Alexander (1990).
Standard for corrosion resistance (extra thickness of steel pylons to
 compensate for unavoidable rust): Norsk Standard, Eurocode 3: Design
 of Steel Structures—Part 5: Piling, NS-EN 1993-5:2007+NA:2010.
Stainless steel cutlery has a life span of at least 100 years: I say this
 because stainless steel cutlery can last "almost forever" and came on the
 market approximately 100 years ago. M. Miodowink, "Stainless Steel
 Revolutionised Eating After Centuries of a Bad Taste in the Mouth,"
 Guardian, April 29, 2015, theguardian.com/technology/2015/apr/29/
 stainless-steel-cutlery-gold-silver-copper-aluminium.

Life span of steel constructions: Sverdrup and Ragnarsdóttir (2014).

CAN WE RUN OUT OF IRON?

Production of iron vs. aluminum; production continues to increase; up to 360 billion metric tons estimated to exist, 30 to 70 billion metric tons already extracted: Sverdrup and Ragnarsdóttir (2014).

On documented reserves and "the lifetime" of the reserves: Arndt et al. (2017).

230 billion metric tons of resources estimated, 83 billion metric tons of reserves, extracting 1.5 billion each year: USGS (2018).

OUT OF THE IRON AGE?

"Quite recently, researchers have looked at all of these mechanisms in context": The system dynamics model analyzes the evolution of production of iron and some other resources in Sverdrup and Ragnarsdóttir (2014).

4 | Copper, Aluminum, and Titanium: From Light Bulbs to Cyborgs

Discussion about when it will be illegal to drive a car yourself, summer 2017: I. E. Fjeld, "Snart blir det ulovlig å kjøre selv" (Soon it will be illegal to drive a car yourself), *NRK,* July 4, 2017, nrk.no/norge/ _-snart-blir-det-ulovlig-a-kjore-selv-1.13581330.

COPPER IN CARS, BODIES, AND WATER

Electric lighting widespread in the 1880s, cheap and reliable electrical energy: Smil (2004).

Amount of copper in cars right after World War II and now: NRC (2008).

Copper in the body: Khurshid and Qureshi (1984).

Copper in water pipes, possible poisoning: Norwegian Institute of Public Health, "Kjemiske og fysiske stoffer i drikkevann" (Chemical and physical substances in drinking water), updated November 19, 2018, fhi. no/nettpub/stoffer-i-drikkevann/ kjemiske-og-fysiske-stoffer-i-drikkevann/kjemiske-og-fysiske-stoffer-i-drikkevann/#kobber-cu.

Copper in metal form used eight thousand years before our time; hammering and processing: Encyclopedia Britannica, "Copper Processing," updated May 1, 2017, britannica.com/technology/copper-processing.

THE COPPER MINES THAT CLEARED THE FORESTS

Extractable deposits in most countries, created via many geological processes, typical concentration of 0.6 percent: Arndt et al. (2017).

Deforestation in Spain, Cyprus, Syria, Iran, Afghanistan: Smil (2004).

Deforestation in Rørosvidda: L. Geithe, "Circumferensen" (Circumference), updated April 7, 2014, bergstaden.org/no/hjem/circumferensen.

Outdoor processing, sulfur in the ore turned into sulfuric acid: L. Geithe, "Kaldrøsting" (Cold calcination), updated September 10, 2013, bergstaden.org/no/kobberverket/smelthytta-pa-roros/kaldrosting.

Copper ore processing done outdoors until the mid-1800s: "Komplex 99139911Malmplassen," regjeringen.no/contentassets/142481976cdc449f964609532920bd68/kompleks_99139911_malmplassen.pdf.

A few decades before production will diminish: Sverdrup and Ragnarsdóttir (2014).

Ten times greater resources if we find deep deposits: Arndt et al. (2017).

ALUMINUM: RED CLOUDS AND WHITE PINES

Most of my electric car is made of aluminum: J. Desjardins, "Extraordinary Raw Materials in a Tesla Model S," *Visual Capitalist*, March 7, 2016, visualcapitalist.com/extraordinary-raw-materials-in-a-tesla-model-s.

Aluminum in the body: Khurshid and Qureshi (1984).

Aluminum in my mobile phone: J. Desjardins, "Extraordinary Raw Materials in an iPhone 6s," *Visual Capitalist*, March 8, 2016, visualcapitalist.com/extraordinary-raw-materials-iphone-6s.

8 percent of Earth's crust is aluminum: Arndt et al. (2017).

Yearly production of iron and aluminum, bauxite extraction, treatment with lye, red mud: Sverdrup and Ragnarsdóttir (2014).

Bauxite extraction from tropical areas (Australia, China, Brazil, and Guinea were the biggest producers in 2017): USGS (2018).

The dam breach in Ajka, ten dead: Wikipedia, "Ajka Alumina Plant Accident," updated June 21, 2018, en.wikipedia.org/wiki/Ajka_alumina_plant_accident.

Limited long-term effects of the Ajka accident: Á. D. Anton et al., "Geochemical Recovery of the Torna-Marcal River System After the Ajka Red Mud Spill, Hungary," *Environmental Science: Processes & Impacts* 16 (2014): 2677–85.

2016 ban on bauxite extraction in Malaysia: USGS (2018); Clean Malaysia, "Bauxite in Malaysia—Will the Ban Bring Relief?," January 26, 2016, cleanmalaysia.com/2016/01/26/bauxite-in-malaysia-will-the-ban-bring-relief.

Aluminum expensive before the end of the 1800s; lowering the melting point with cryolite; electrical circuits (the Hall-Heroult Process): Street and Alexander (1990).

The aluminum plant in Årdal, history: Industrimuseum, "Årdal og Sundal Verk A/S," industrimuseum.no/bedrifter/aardalogsundalverka_s.

Norway, the world's eighth-largest producer of aluminum: USGS (2018).

Damage to livestock, the start of Norwegian environmental politics: K. Tvedt, "Bakgrunn: Forgiftet fe ga norsk miljøpolitikk" (Background: poisoned livestock started Norwegian environmental politics), forskning. no, January 23, 2012, forskning.no/ husdyr-moderne-historie-miljopolitikk/2012/01/ forgiftet-fe-ga-norsk-miljopolitikk.

Purification systems in the 1980s: Wikipedia, "Årdal," updated June 6, 2018, no.wikipedia.org/wiki/Årdal.

Continued effects on deer teeth: O. R. Sælthun, "Mykje fluorskader på hjorten i Årdal" (A great deal of fluoride damage to deer in Årdal), Porten.no, February 22, 2017, porten.no/artiklar/ mykje-fluorskader-pa-hjorten-i-ardal/393074; O. R. Sælthun, "Hydro: Vanskeleg å forstå at resultata er slik" (Hydro: Difficult to understand that the results are like this), Porten.no, February 22, 2018, porten. no/artiklar/hydro-vanskeleg-a-forsta-at-resultata-er-slik/393079; Norwegian Veterinary Institute, "Helseovervåkingsprogrammet for hjortevilt og moskus (HOP) 2017" (Monitoring Program for Deer and Musk), www.vetinst.no/rapporter-og-publikasjoner/rapporter/2018/ helseovervakingsprogrammet-for-hjortevilt-og-moskus-hop-2017.

USING WHAT WE'VE ALREADY USED

Aluminum from other minerals; 60 percent recycling of aluminum likely more important than mining in a few decades: Sverdrup and Ragnarsdóttir (2014).

Elements in mobile phones: Desjardins, "Extraordinary Raw Materials in an iPhone 6s."

THE TITANIUM IN A MOUNTAIN

My car's undercarriage is made of titanium: Desjardins, "Extraordinary Raw Materials in a Tesla Model S."

Titanium in the human body, artificial teeth made of cast iron, artificial hips in 1938, need for materials in implants: Giselbrecht et al. (2013).

90 percent is used for pigment; some of the world's largest deposits in solid rock: Norwegian Institute for Cultural Heritage Research (NIKU), Konsekvensutredning for utvinning av rutil i Engebøfjellet, Naustdal kommune (Impact assessment for rutile extraction in Engebøfjellet, Naustdal municipality), Landscape Department report 30/08.

Extracted titanium in Norway for over a hundred years (extraction from the Kragerø field at the beginning of the twentieth century): Store Norske Leksikon, "Norsk bergindustrihistorie," (Norwegian rock industry history), updated December 20, 2016, snl.no/ Norsk_bergindustrihistorie.

Titanium from sand: Gilchrist (1989).

Using magnets, gravity, and foam (flotation) to sort out titanium-containing minerals, and environmental impacts of sea deposits: Norwegian Climate and Pollution Agency (Klif), Gruvedrift i Engebøfjellet—Klifs vurdering og anbefaling (Mining in Engebøfjellet—Klif's assessment and recommendation), March 19, 2012.

On the battle against sea deposits and the geochemical arguments for sea deposits vs. landfill, two million metric tons of sludge per year in Titania Landfill: P. Aagaard and K. Bjørlykke, "Naturvernere lager naturkatastrofe" (Nature conservation creates environmental disaster), forskning.no, June 14, 2017, forskning.no/ naturvern-geofag-stub/2008/02/naturvernere-lager-naturkatastrofe.

After thirty years, the Jøssingfjord is still showing signs of impact from the deposits: L. M. Kalstad et al., "Urovekkende funn på bunnen av Jøssingfjorden" (Disturbing discoveries at the bottom of the Jøssingfjord), *NRK*, May 25, 2017, nrk.no/rogaland/ urovekkende-funn-pa-bunnen-av-jossingfjorden-1.13532071.

THE CYBORGS ARE COMING!

Most of the information in this section is taken from Giselbrecht et al. (2013).

Chips in hands at American workplaces: O. Ording, "Låser opp dører med en chip under huden" (Unlocking doors with a chip under the skin), NRK, August 13, 2017, nrk.no/norge/ laser-opp-dorer-med-en-chip-under-huden-1.13637732.

Arne Larsson, first pacemaker: L. K. Altman, "Arne H. W. Larsson, 86; Had First Internal Pacemaker," *The New York Times*, January 18, 2002, nytimes.com/2002/01/18/world/ arne-h-w-larsson-86-had-first-internal-pacemaker.html.

THE FUTURE OF MACHINE PEOPLE

In the future, machines in the body could operate without batteries: Giselbrecht et al. (2013).

Considerable requirements for systems that make perfectly clean components: E. D. Williams et al., "The 1.7 Kilogram Microchip: Energy and Material Use in the Production of Semiconductor Devices," *Environmental Science and Technology* 36 (2002): 5504–10.

Chemical separation requires one third as much energy as the entire transport sector: The transport sector makes up 35 percent of the world's energy consumption: International Energy Agency, Key World Energy Statistics 2017, doi.org/10.1787/key_energ_stat-2017-en.

Chemical separation makes up 10 to 15 percent of the world's energy consumption: D. S. Sholl and R. P. Lively, "Seven Chemical Separations to Change the World," *Nature* 532 (2016): 435–37.

Bacteria that make nanotubes: Y. Tan et al., "Expressing the Geobacter metallireducens PilA in Geobacter sulfurreducens Yields Pili with Exceptional Conductivity," *mBio* 8 (2017): e02203–16.

Requirements for materials to be used in space: W. Wassmer, "The Materials Used in Artificial Satellites and Space Structures," Azo Materials, May 12, 2015, azom.com/article.aspx?ArticleID=12034.

5 | Calcium and Silicon in Bones and Concrete

Teeth and bones contain calcium, phosphorus, and oxygen, as well as silicon (osteoblasts—cells that make bone tissue—contain silicon): Khurshid and Qureshi (1984).

HARD AND BRITTLE

Ceramic materials, definition and properties: B. Basu and K. Balani, *Advanced Structural Ceramics*, Wiley, 2017.

MOLDING WITH CLAY

Technical definition of clay: Wikipedia, "Clay," updated February 21, 2020, en.wikipedia.org/wiki/Clay.

Crystal structure, clay minerals: James Hutton Institute, "Clay Minerals," claysandminerals.com/minerals/clayminerals.

Ceramics production, history: American Ceramic Society, "A Brief History of Ceramics and Glass," ceramics.org/about/what-are-engineered-ceramics-and-glass/brief-history-of-ceramics-and-glass.

Porcelain, developed by the Chinese in the seventh century: Encyclopaedia Britannica, "Porcelain," updated January 10, 2020, britannica.com/art/porcelain.

THE MESSY ATOMS IN THE WINDOWPANE

Oldest human-made glass 4,500 years old: S. C. Rasmussen, *How Glass Changed the World*, Springer, 2012.

Glass in volcanoes, earthquakes, meteorite strikes: B. P. Glass, "Glass: The Geologic Connection," International Journal of Applied Glass Science 7 (2016): 435–45.

Contents of glass, how glass is produced; just a small amount of the wrong glass in the furnace could be enough to have to throw everything away: L. L. Gaines and M. M. Mintz, *Energy Implications of Glass Container Recycling*, US Department of Energy Report ANL/EDS-18 NREL/TP-430-5703, osti.gov/servlets/purl/10161731.

Manufacturing of glass, molds, windows: Safeglass (Europe) Limited, "Modern Glass Making Techniques," breakglass.org/Glass_making.html.

The glass in the windshield is cooled down extra quickly: Wikipedia, "Tempered Glass," updated February 1, 2020, en.wikipedia.org/wiki/Tempered_glass.

Refractory mold with boron oxide: Wikipedia, "Borosilicate Glass," updated January 13, 2020, en.wikipedia.org/wiki/Borosilicate_glass.

Crystal glass with lead, and is it dangerous to drink from a crystal glass that contains lead?: Wikipedia, "Lead Glass," updated February 8, 2020, en.wikipedia.org/wiki/Lead_glass.

Glass in advanced communication, more important in the future: NRC (2018).

FROM ALGAE TO CONCRETE

Flint deposits in the Nordics: Store Norske Leksikon, "Flint—arkeologi," updated October 26, 2018, snl.no/Flint_-_arkeologisk.

Limestone is broken down at over 800°C (but for efficiency, the furnace often must be heated to much higher temperatures): B. R. Stanmore and P. Gilot, "Review—Calcination and Carbonation of Limestone During Thermal Cycling for CO_2 Sequestration," Fuel Processing Technology 86 (2005): 1707–43.

About calcination, slaked lime, lime mortar, and its early use: Courland (2011).

VOLCANIC ASH IN THE COLOSSEUM

Most of the information in this section is taken from Courland (2011).

Volcanic eruption, Santorini, 1640 BCE (a search of recent articles will show that the exact date is still being debated): T. Pfeiffer, "Vent Development During the Minoan Eruption (1640 BC) of Santorini, Greece, as Suggested by Ballistic Blocks," *Journal of Volcanology and Geothermal Research* 106 (2001): 229–42.

The volcanic eruption and the subsequent tsunami led to the fall of Minoan culture: This is a leading hypothesis, but not universal. See, for example, J. Grattan, "Aspects of Armageddon: An Exploration of the Role of Volcanic Eruptions in Human History and Civilization," *Quaternary International* 151 (2006): 10–18.

CONCRETE THAT SCRAPES THE CLOUDS

The information in this section is taken from Courland (2011). In addition, I've used my own experience from several years of research on concrete and materials.

IS THERE ENOUGH SAND?

Until recently, sand and gravel were collected near the construction site; concrete production quadrupled in China and increased 50 percent in the rest of the world over the past twenty years; few natural sand deposits remain in Europe: H. U. Sverdrup et al., "A Simple System Dynamics Model for the Global Production Rate of Sand, Gravel, Crushed Rock and Stone, Market Prices and Long-Term Supply Embedded into the WORLD6 Model," *BioPhysical Economics and Resource Quality* 2 (2017): 8.

Suitability of various types of sand and gravel in concrete; 70 to 90 percent of solid material extracted, 180 million metric tons for industry, twice as much as the world's rivers; effect of withdrawals on rivers and oceans; projects in Dubai and Singapore: United Nations Environment Programme, "Sand, Rarer Than One Thinks," *Global Environment Alert Service,* March 2014, hdl.handle.net/ 20.500.11822/8665.

Twice as much concrete as all other building materials: C. R. Gagg, "Cement and Concrete as an Engineering Material: An Historical Appraisal and Case Study Analysis," *Engineering Failure Analysis* 40 (2014): 114–40.

LIVING CERAMICS FACTORIES

Sea urchin, mother-of-pearl are strong materials; research on creating such materials: N. A. J. M. Sommerdijk and G. de With, "Biomimetic $CaCO_3$ Mineralization Using Designer Molecules and Interfaces," *Chemical Reviews* 108 (2008): 4499–550.

Bacterial concrete, use of biotechnology in construction materials: V. Stabnikov et al., "Construction Biotechnology: A New Area of Biotechnological Research and Applications," *World Journal of Microbiology and Biotechnology* 93 (2015): 1224–35.

6 | Multitalented Carbon: Nails, Rubber, and Plastic

Information on how things were done in hospitals before we had plastic: I found this discussion on a nurses' website: sunnyjohn, "What was IV tubing made of before the invention of plastics?," General Nursing forum, allnurses.com, August 5, 2005, allnurses.com/ what-iv-tubing-made-invention-t84632.

About the history of blood banks and the importance of plastic blood bags: C. W. Walter, "Invention and Development of the Blood Bag," *Vox Sanguinis* 47 (1984): 318–24.

NATURAL RUBBER AND VENERABLE VULCANIZATION

About natural rubber and vulcanization: Massy (2017).

Glass, sealed with rubber gaskets: L. Meredith, "The Brief History of Canning Foods," The Spruce Eats, updated October 2, 2019, thespruceeats.com/brief-history-of-canning-food-1327429.

Rubber extraction in the Congo: A. Hochschild, *King Leopold's Legacy,* Pax Publishers, 2005.

The structure of keratin: Wikipedia, "Keratin," updated June 29, 2018, en.wikipedia.org/wiki/Keratin.

FROM TIMBER TO TEXTILES

The structure of cellulose and its materials: Massy (2017).

PLASTICS OF THE PAST

For a discussion on whether plastic reduces food waste, see J.-P. Schwetizer et al., *Unwrapped: How Throwaway Plastic Is Failing to Reduce Europe's Food Waste Problem (And What We Need to Do Instead),* Institute for European Environmental Policy (IEEP), 2018.

Solid materials become liquid at about two miles deep (two to four kilometers); dinosaurs and trees turn into coal; algae and other small creatures can turn into oil: S. Chernicoff and H. A. Fox, *Essentials of Geology,* 2nd ed., Houghton Mifflin, 2000.

Leo Baekland, first plastic made from fossil sources: Massy (2017); J. Jiang and N. King, "How Fossil Fuels Helped a Chemist Launch the Plastic Industry," September 29, 2016, All Things Considered, transcript and audio at Planet Money, npr.org/2016/09/29/495965233/how-fossil-fuels-helped-a-chemist-launch-the-plastic-industry?t=1530770723354.

New materials and their use, plastic additives: Massy (2017).

Nearly 400 million metric tons of plastic today (380 million metric tons in 2015): R. Geyer et al., "Production, Use and Fate of All Plastics Ever Made," *Science Advances* 3 (2017): e1700782.

Oil consumption today is 4 billion metric tons: OECD, "Crude Oil Production (Indicator)," accessed July 5, 2018, doi.org/10.1787/4747b431-en.

THE TRASH ISLAND

The study of plastic on Henderson Island: Lavers and Bond (2017).

WHAT DO WE DO WITH ALL THIS PLASTIC?

The origin of plastic on Henderson Island: Lavers and Bond (2017).

Whale with over forty plastic bags in its stomach: Store Norske Leksikon, "Plasthvalen" (Plastic whale), updated November 2, 2017, snl.no/plasthvalen.

Plastic waste in our own bodies: A. D. Vethaak and H. A. Leslie, "Plastic Debris Is a Human Health Issue," *Environmental Science and Technology* 50 (2016): 6825–26.

What happens to the plastic in the blue bags: "Hva skjer med plasten?" (What happens to the plastic?), Esval Miljøpark (Esval environmental park), esval.no/renovasjon/kildesortering/hva_skjer_med_plasten_.

Plastic is not very suitable for recycling: Massy (2017).

Burning plastic safely: A. Herring, "Burning Plastic as Cleanly as Natural Gas," phys.org, December 5, 2013, phys.org/news/ 2013-12-plastic-cleanly-natural-gas.html.

Plastic that can be broken down by microorganisms: V. Piemonte, "Inside the Bioplastics World: An Alternative to Petroleum-Based Plastics," in *Sustainable Development in Chemical Engineering—Innovative Technologies*, ed. V. Piemonte, John Wiley & Sons (2013); OECD (2011).

PLASTIC AFTER OIL

Projected that plastic production will increase to one billion metric tons annually; use of organisms living under extreme conditions; changing genes: OECD (2011).

The first Lego blocks were made of cellulose: K. Heggdal and C. Veløy, "Fremtidens klimavennlige Lego-univers" (The climate-friendly Lego universe of the future), *NRK*, December 3, 2015, nrk.no/viten/xl/ fremtidens-klimavenn-lige-lego-univers-1.12679556.

Use of cellulose, chitin, lignin, plant oils, lactic acid, bacteria that make cellulose fibers: A. Gandini, "Polymers from Renewable Resources: A Challenge for the Future of Macromolecular Material," *Macromolecules* 41 (2008): 9491–504.

7 | Potassium, Nitrogen, and Phosphorus: The Elements That Give Us Food

THE JOURNEY TO THE DEAD SEA

Pumping plants in the 1960s and '70s, the extraction plant in the southern part, production of carnallite: Holmes (2010).

120 feet lower than before the pumping stations: S. Griffiths, "Slow Death of the Dead Sea: Levels of Salt Water Are Dropping by One Meter Every Year," MailOnline, January 5, 2015, dailymail.co.uk/sciencetech/article-2897538/ Slow-death-Dead-Sea-Levels- salt-water-dropping-one-metre-year.html.

The water level dropped by 3 feet per year: Israel Oceanographic & Limnological Research, "Long-Term Changes in the Dead Sea," isramar.ocean.org.il/isramar2009/DeadSea/LongTerm.aspx.

NUTRIENTS IN OUR NERVES

Potassium's function in the body: Khurshid and Qureshi (1984).

POTASSIUM FROM WATER

Potassium extraction: The Canadian Encyclopedia, "Potash," updated March 4, 2015, thecanadianencyclopedia.ca/en/article/potash.

The world's largest potassium producers; reserves and resources: USGS (2018).

Groundwater resources are in the process of being emptied: C. Dalin et al., "Groundwater Depletion Embedded in International Food Trade," *Nature* 543 (2017): 700–704.

NITROGEN FROM AIR

Nitrogen makes up 3.2 percent of human body weight: Wikipedia, "Composition of the Human Body," updated July 2, 2018, en.wikipedia.org/wiki/Composition_of_ the_human_body.

Nitrogen in the atmosphere, transformation into forms that can be absorbed by plants, the nitrogen cycle: A. Appelo and D. Postma, *Geochemistry, Groundwater and Pollution*, 2nd ed., A.A. Balkema, 2005.

The Birkeland-Eyde process: Wikipedia, "Birkeland-Eyde Process," updated January 31, 2020, en.wikipedia.org/wiki/Birkeland–Eyde_process.

Laying the foundation for Norsk Hydro's production of artificial fertilizers, and the transition to the Haber-Bosch process: Wikipedia, "Norsk Hydro," updated January 31, 2020, en.wikipedia.org/wiki/Norsk_Hydro.

The Haber-Bosch process: Wikipedia, "Haber Process," updated January 8, 2020, en.wikipedia.org/wiki/Haber_process.

Half of the nitrogen in agriculture comes from fertilizers; enough nitrogen fertilizer for a thousand years with all known natural gas reserves; alternative production methods: M. Blanco, Supply of and Access to Key Nutrients NPK for Fertilizers for Feeding the World in 2050, ETSI Agrónomos UPM, November 28, 2011.

Making the Birkeland-Eyde process more energy efficient: O. R. Valmot, "Vil kapre enormt marked med over 100 år gammel norsk teknologi," Teknisk Ukeblad, January 28, 2016, tu.no/artikler/ vil-kapre-enormt-marked-med-over-100-ar-gammel-norsk-teknologi/276467.

Genetic modification of nitrogen fixation: F. Mus et al., "Symbiotic Nitrogen Fixation and the Challenges of Its Extension to Nonlegumes," *Applied and Environmental Microbiology* 82 (2016): 3698–710.

PHOSPHORUS FROM ROCKS

Phosphorus in solid form or stuck to mineral surfaces: Appelo and Postma, Geochemistry, Groundwater and Pollution.

Phosphorus is 1 percent of human body weight: Wikipedia, "Composition of the Human Body.

Historical use of phosphorus fertilizer, amount of geological phosphorus used today, use in organic farming, 20 percent of extracted phosphorus reaches food, loss of phosphorus, proportion of nutrients returned to the soil, methods of reducing geological phosphorus dependence: Cordell et al. (2009).

Major producers, resources, and reserves: USGS (2018).

Morocco and Western Sahara: A. Kasprak, "The Desert Rock That Feeds the World," *Atlantic,* November 29, 2016, theatlantic.com/science/archive/2016/11/the-desert-rock-that-feeds-the-world/508853.

Extraction from the seabed outside New Zealand: Chatham Rock Phosphate, "The Project Overview," rockphosphate.co.nz/the-project. The permit application was rejected in 2015, but the company is trying again. See R. Howard, "Chatham Rock Says Rejection of EPA Cost Claim Will Hurt Cash Flow," *National Business Review,* December 12, 2017, nbr.co.nz/article/chatham-rock-says-rejection-epa-costs-claim-will-hurt-cash-flow-b-211056.

Extraction from the seabed outside Namibia (Sandpiper Phosphate), permission in 2016: E. Smit, "Phosphate Mining Gets Green Light," Ministry of Environment and Tourism Namibia, October 19, 2016, met.gov.na/news/159/phosphate-mining-gets-green-light. The permit was later revoked following disagreement with local groups, and the case is not yet closed: G. Mathope, "Marine Phosphate Mining Gets Namibians Hot Under the Collar," *Citizen,* April 26, 2017, citizen.co.za/business/1497708/marine-phosphate-mining-gets-namibians-hot-collar.

Warning of dramatic lack of phosphorus for food production in less than 100 years: Cordell et al. (2009); Sverdrup and Ragnarsdóttir (2014).

Documented reserves, can extract phosphorus for over 1,100 years: USGS (2018) states total resources are around 300 billion metric tons, 263 million metric tons were mined in 2017, 300 billion metric tons / 263 million metric tons per year = 1,100 years. For discussion on whether we will observe scarcity of phosphorus in a few decades, see also: F.-W. Wellmer, "Discovery and Sustainability," in *Non-Renewable Resources Issues,* ed. R. Sinding-Larsen and F.-W. Wellmer, Springer, 2012; and R. W. Scholz and F.-W. Wellmer, "Approaching a Dynamic View on the Availability of Mineral Resources: What May We Learn from the Phosphorous Case?," *Global Environmental Change* 23 (2012): 11–27.

Nature spends a hundred years creating one inch of topsoil, loss of topsoil occurs ten to a hundred times faster than new soil is formed, today phosphorus loss is six times greater than the natural supply, optimizing agriculture to get the most phosphorus from the environment, population reduction: Sverdrup and Ragnarsdóttir (2014).

The Dust Bowl catastrophe: Wikipedia, "Dust Bowl," updated July 8, 2018, en.wikipedia.org/wiki/Dust_Bowl.

Half of the topsoil in the Midwest lost in the last hundred years: K. W. Butzer, "Accelerated Soil Erosion: A Problem of Man-Land Relationships," in *Perspectives on Environment*, ed. I. R. Manners and M. W. Mikesell, Association of American Geographers, 1974.

Erosion in the Dead Sea: E. Oddone, "The Death of the Dead Sea," NOVA Next, August 17, 2016, pbs.org/wgbh/nova/article/dead-sea-dying.

Strategies for preventing erosion from agricultural land: Pipkin (2005).

Urine-separating toilets in Sweden: Sweden Water and Sewer Guide, "Toilets," https://avloppsguiden.se/informationssidor/toaletter.

NUTRIENTS GONE ASTRAY

Little nutrition is returned, reasons: J. M. McDonald et al., Manure Use for Fertilizer and for Energy: Report to Congress US Department of Agriculture, 2009.

Computerized agricultural machinery to supply the exact amount of fertilizer needed: Norsk Landbrukssamvirke, "Presisjonslandbruket vil redusere klimagassutslipp" (Precision agriculture will reduce greenhouse gas emissions), updated September 18, 2018, landbruk.no/biookonomi/presisjonslandbruk-redusere-klimagassutslipp.

THE FUTURE OF THE DEAD SEA

The bottom of the evaporation pools rises by almost seven inches a year (17.8 cm): Holmes (2010).

8 | Without Energy, Nothing Happens

ENERGY FROM THE SUN

Reflections on how human energy consumption has evolved differently from animals' are taken from Smil (2004).

DRAINING EARTH'S ENERGY STORES

Stored energy decreased two-thirds in 1900 compared to year 0; today about half remains (55 percent remained in the year 2000; I have assumed continued reduction); 85 percent of today's energy comes from fossil energy sources; the amount of energy to civilization is one quarter of the energy plants capture from the sun: J. R. Schramski et al., "Human Domination of the Biosphere: Rapid Discharge of the Earth-Space Battery Foretells the Future of Mankind, *PNAS* 112 (2015): 9511–17.

World population, 2012: US Census Bureau, International Database, "Total Midyear Population for the World: 1950–2050," web.archive.org/web/20120121175120/http://www.census.gov/population/international/data/idb/worldpoptotal.php, updated June 27, 2011; 7.6 billion in 2018: Worldometer, worldometers.info/world-population.

The energy supply for my family compared to before: Smil (2004); S. Arneson, "En norsk husholdning har samme energi- bruk som 3000 slaver og 200 trekkdyr" (A Norwegian household uses the same amount of energy as 3,000 slaves and 200 migratory animals), Teknisk Ukeblad, January 13, 2015, tu.no/artikler/kommentar-en-norsk-husholdning-har-samme-energiforbruk-som-3000-slaver-og-200-trekkdyr/223656. The estimates from Smil (USA) have been adjusted down to adapt to Norwegian conditions.

THE SOCIETY WE WANT

Energy surplus, specialization, and proportion of the population in food production: R. Heinberg, *Peak Everything: Waking Up to the Century of Declines*, New Society Publishers, 2007.

The order of priority tasks ("Pyramid of energetic needs"): J. G. Lambert et al., "Energy, EROI and Quality of Life," *Energy Policy* 64 (2014): 153–67.

ENERGY IN, ENERGY OUT

This section is based on the Energy Return on Investment (EROI) concept, also known as Energy Return on Energy Invested (EROEI). EROI = energy out / energy in.

EROI = 20 for a good life, EROI > 10 for industrial society, EROI = 3 minimum target for primitive civilization, EROI = 10 for hunters and gatherers, EROI = 100 for oil extracted in the 1930s: C. A. S. Hall et al., "What Is the Minimum EROI That a Sustainable Society Must Have?," *Energies* 2 (2009): 25–47.

EROI = 20 for today's conventional oil field (world average 18 in 2005): C. A. S. Hall et al., "EROI for Different Fuels and the Implications for Society," *Energy Policy* 64 (2014): 141–52.

EROI = 10 (under 10) for unconventional oil sources: D. J. Murphy, "The Implications of the Declining Energy Return on Investment of Oil Production," Philosophical Transactions of the Royal Society A: *Mathematical, Physical, and Engineering Sciences*, 327 (2014): 20130136.

OUT OF THE FOSSIL SOCIETY

Most people agree that we've spent significant amounts of fossil energy resources; the oil age will end during this century or the next: Resources will last 80 to 240 years with today's consumption: Sverdrup and Ragnarsdóttir (2014).

Climate change and consequences: Intergovernmental Panel on Climate Change (IPCC), *Climate Change* 2014: Synthesis Report, IPCC, 2014, ipcc.ch/report/ar5/syr.

GEOTHERMAL HEAT AND NUCLEAR POWER: ENERGY FROM EARTH'S BEGINNINGS

Radioactive materials when neutron stars collide with each other or with black holes: S. Rosswog, "Viewpoint: Out of Neutron Star Rubble Comes Gold," *Physics,* December 6, 2017, physics.aps.org/articles/v10/131.

Heat flow is generally too small but can be useful in specific places: D. J. C. McKay, *Sustainable Energy—Without the Hot Air,* UIT Cambridge Ltd, 2009.

Iceland large aluminum producer (Iceland was the world's tenth-largest aluminum producer in 2017): USGS (2018).

How the reactor in a nuclear power plant works, chain reaction: Store Norske Leksikon, "Kjernereaktor" (Nuclear reactor), updated January 21, 2015, snl.no/kjernereaktor.

Out of uranium in 60 to 140 years with today's technology; new technology could power us 25,000 years into the future: Sverdrup and Ragnarsdóttir (2014).

New reactors, high demands on materials in new reactors, and nuclear risk: See F. Pearce, "Are Fast-Breeder Reactors the Answer to Our Nuclear Waste Nightmare?," *Guardian,* July 30, 2012, theguardian.com/environment/2012/jul/30/fast-breeder-reactors-nuclear-waste-nightmare, and N. Touran, "Molten Salt Reactors," WhatIsNuclear, whatisnuclear.com/msr.html.

POWER STRAIGHT FROM THE SUN

How solar cells work: Wikipedia, "Solar Cells," accessed February 21, 2020, en.wikipedia.org/wiki/Solar_cell.

Solar cells accelerating development; many believe they will be the most important: International Energy Agency (IEA), World Energy Outlook 2017, IEA, 2017, iea.org/weo2017.

Supplies of lead will decline before 2050, followed by tin and silver: Sverdrup and Ragnarsdóttir (2014).

Contents of new types of solar cells (gallium, tellurium, indium, selenium) and pigment solar cells: Öhrlund (2011).

Selenium linked to copper, gallium to aluminum: V. Steinbach and F.-W. Wellmer, "Consumption and Use of Non-Renewable Mineral and Energy Raw Materials from an Economic Geology Point of View," *Sustainability* 2 (2010): 1408–30.

About solar cells with dyes: Wikipedia, "Dye-Sensitized Solar Cell," updated July 9, 2018, en.wikipedia.org/wiki/Dye-sensitized_solar_cell.

WATER THAT RUNS, WIND THAT BLOWS

Greater potential for development of wind power than hydropower: See figures from the International Energy Agency, "Hydropower," iea.org/topics/renewables/hydropower (119 GW increase 2017–2022), and "Wind," iea.org/topics/renewables/wind (295 GW increase 2017–2022), both parts from IEA, Renewables 2017, IEA, 2017.

Development of wind turbines accelerated in the 1970s; life expectancy of wind turbines twenty to thirty plus fifteen years; life span of coal and nuclear power plants thirty to fifty years; materials in wind turbines: Wilburn (2011).

Could have produced all the energy we need in Norway with wind: Net energy consumption in Norway 2016 was 214 terawatt hours (TWh): Statistics Norway, "Production and Consumption of Energy, Energy Balance," updated June 20, 2018, ssb.no/energi-og-industri/statistikker/energibalanse. Theoretical resource basis for onshore wind power is 1,400 TWh: Norwegian Water Resources and Energy Directorate (NVE), "Resource Base," updated April 11, 2019, nve.no/energiforsyning/ressursgrunnlag.

Hydropower EROI > 100, wind power EROI = 20: Hall et al., "EROI for Different Fuels and the Implications for Society."

THE RARE-EARTH ELEMENTS

On rare-earth elements: Wikipedia, "Rare-Earth Element," accessed February 21, 2020, en.wikipedia.org/wiki/Rare-earth_element.

Neodymium and sixty other elements in the mobile phone: J. Desjardins, "Extraordinary Raw Materials in an iPhone 6s," *Visual Capitalist*, March 8, 2016, visualcapitalist.com/extraordinary-raw- materials-iphone-6s.

China dominates production; Brazil has next-largest known reserves: USGS (2018).

Challenges in the development of wind turbines: Öhrlund (2011); Wilbur (2011).

The Fen Complex, perhaps Europe's largest deposit; mapping today: J. Seehusen, "Norge kan sitte på Europas største forekomst av sjeldne jordarter" (Norway could be sitting on Europe's largest occurrence of rare-earth elements) Teknisk Ukeblad, July 23, 2017, tu.no/artikler/norge-kan-sitte-pa-europas-storste-forekomst-av-sjeldne-jordarter/398067.

The Fen Complex, geological history: S. Dahlgren, "Fensfeltet—et stykke eksplosiv geologi" (The Fen Complex—A piece of explosive geology), *Stein magasin for populærgeologi* 3 (1993): 146–55.

POWER ON A QUIET WINTER NIGHT

Pump power plants: Wikipedia, "Pumped-Storage Hydroelectricity," updated February 19, 2020, en.wikipedia.org/wiki/Pumped-storage_hydroelectricity.

Energy stored in flywheels: S. Springborg, "Danskere vil opfinde svinghjul til lagring af vind-og solenergi" (Danes want to invent flywheels for storing wind and solar energy), EnergiWatch, November 1, 2017, energiwatch.dk/Energinyt/Cleantech/article9993112.ece.

Energy stored in molten salt: Wikipedia, "Thermal Energy Storage: Molten-salt technology," updated February 7, 2020, en.wikipedia.org/wiki/Thermal_energy_storage#Molten-salt_technology.

COBALT IN THE BATTERY

Use of lithium and cobalt in batteries: Wikipedia, "Lithium-Ion Battery," updated July 18, 2018, en.wikipedia.org/wiki/Lithium-ion_battery.

Extraction from solid rock (spodums) in Australia and saltwater sources (brine) in Argentina and Chile; estimated resources over 1,200 years with current production (total resources 53 million metric tons, production in 2017 at 43,000 metric tons, yields 1,233 years with current production): USGS (2018).

Extraction of cobalt in the Congo: USGS (2018); T. C. Frankel, "The Cobalt Pipeline: Tracing the Path from Deadly Hand-Dug Mines in Congo to Consumers' Phones and Laptops," *Washington Post*, September 30, 2016, washingtonpost.com/graphics/business/batteries/congo-cobalt-mining-for-lithium-ion-battery.

Energy density in oil (approx. 55 MJ/kg), lithium-ion batteries (theoretical maximum approx. 3 MJ/kg), lithium-air batteries (theoretical maximum approx. 43 MJ/kg); one kilogram of hydrogen has three times as much energy as one kilogram of oil: K. Z. House and A. Johnson, "The Limits of Energy Storage Technology," *Bulletin of the Atomic Scientists*, January 20, 2009, thebulletin.org/2009/01/the-limits-of-energy-storage-technology.

Fuel cells contain platinum: International Platinum Group Metals Association, "Fuel Cells," ipa-news.de/index/pgm-applications/automotive/fuel-cells.html.

South Africa the largest producer of platinum; four other countries: USGS (2018).

Platinum among one of the elements that the authorities pay extra attention to: NRC (2008).

GASOLINE FROM PLANTS

The function of organic matter in soil, how biofuels are produced: A. Friedemann, "Peak Soil: Why Cellulosic Ethanol, Biofuels Are Unsustainable and a Threat to America," resilience.org, April 13, 2007, resilience.org/stories/2007-04-13/peak-soil-why-cellulosic-ethanol-biofuels-are-unsustainable-and-threat-america.

Photosynthesis stores a maximum of 12 percent of solar energy; can get EROI = 50 for energy-rich plants and lots of sun; biofuels on the market today have an EROI between 2 and 5, around 1 for difficult resources: A. K. Ringsmuth et al., "Can Photosynthesis Enable a Global Transition from Fossil Fuels to Solar Fuels, to Mitigate Climate Change and Fuel-Supply Limitations?," *Renewable and Sustainable Energy Reviews* 62 (2016): 134–63.

Solar cells could store about 20 percent of solar energy: Wikipedia, "Solar Cells," accessed February 21, 2020, en.wikipedia.org/wiki/Solar_cell.

Growing algae in tanks or pipes: OECD (2011).

TODAY WE EAT OIL

The phrase "today we eat oil"; ten times more energy needed to produce food than it provides; increased food production since the 1950s: D. A. Pfeiffer, "Eating Fossil Fuels," resilience.org, October 2, 2003, resilience.org/stories/2003-10-02/eating-fossil-fuels.

9 | Plan B

UNLIMITED ENERGY: A SUN ON EARTH

Temperature and pressure in the sun; research during the Cold War; tokamak and stellarators—design and challenges; stellarator first proposed in the 1950s, developed in the 1980s: A. Mann, "Core Concepts: Stabilizing Turbulence in Fusion Stellarators," *PNAS* 114 (2017): 1217–19.

Use of deuterium and tritium; today tritium is produced by a rare variant of lithium; a thousand years of energy consumption with lithium from Earth's crust, several million years with sea extraction; no meltdowns or explosions in the fusion reactor; plasma is trapped by magnetic fields; radioactive material produced in fusion reactor: S. C. Cowley, "The Quest for Fusion Power," *Nature Physics* 12 (2016): 384–86.

The sabotage action against Norsk Hydro: Wikipedia, "Norwegian Heavy Water Sabotage," accessed February 21, 2020, en.wikipedia.org/wiki/Norwegian_heavy_water_sabotage.

ITER first plasma in 2025: ITER, "Building ITER," iter.org/construction/construction.

Wendelstein 7-X Plasma in 2016: Max Planck Institute for Plasma Physics, "Wendelstein 7-X: Upgrading After Successful First Round of Experiments," press release, July 8, 2016, ipp.mpg.de/4073918/07_16.

ELEMENTS IN SPACE

Tutankhamen's dagger: Comelli et al. (2016).

Thousands of meteorites every year; 2,500 metric tons of iron, 600 metric tons of nickel, 100 metric tons of cobalt: Sverdrup and Ragnarsdóttir (2014).

Extract 1.5 billion metric tons of iron, 2 million metric tons of nickel, 110,000 metric tons of cobalt: USGS (2018).

Thousands of known asteroids in the solar system; asteroid belts 90 to 280 million miles from Earth (185 to 370 million miles from the Sun, Earth is 95 million miles from the Sun); Ceres largest with 600-mile diameter; likely several thousand nearer Earth, 250 known, several thousand could hit Earth; contents of asteroids: NASA, "Near Earth Rendezvous (NEAR) Press Kit," February 1996, https://www.nasa.gov/home/hqnews/presskit/1996/NEAR_Press_Kit/NEARpk.txt.

How asteroid mining will take place: Science Clarified, "How Humans Will Mine Asteroids and Comets," scienceclarified.com/scitech/Comets-and-Asteroids/How-Humans-Will-Mine-Asteroids-and-Comets.html.

Effect of long-term weightlessness on the body (describing the most recent findings of the "twin study," in which Scott Kelly spent almost a year on the International Space Station while his twin brother was on Earth; more results will come later in 2018): J. Parks, "How Does Space Change the Human Body?," *Astronomy*, February 16, 2018, astronomy.com/news/2018/02/how-does-space-change-the-human-body.

Hayabusha: E. Howell, "Hayabusha: Troubled Sample-Return Mission," Space.com, March 30, 2018, space.com/40156-hayabusa.html.

Hayabusha 2: E. Howell, "Hayabusha2: Japan's 2nd Asteroid Sample Mission," Space.com, July 9, 2018, space.com/40161-hayabusa2.html.

OSIRIS-REx: NASA, "About OSIRIS-REx," nasa.gov/mission_pages/
osiris-rex/about. According to NASA's "Mission Status" page, everything
is going according to plan and the vessel is currently orbiting Bennu in
preparation for sample collection: asteroidmission.org/status-updates.

Commercial companies with mining plans, focus on resources that are
useful in space: C. P. Persson, "Gruvedrift på asteroider: Første skritt
blir drivstoffstasjoner i verdensrommet (Mining on asteroids: The
first step will be fuel stations in space), forskning.no, April 8, 2017,
forskning.no/romfart/
gruvedrift-pa-asteroider-forste-skritt-blir-drivstoffstasjoner-i-
verdensrommet/
354247.

AWAY FROM EARTH?

Elon Musk on leaving Earth: M. Mosher and K. Dickerson, "Elon Musk:
We Need to Leave Earth as Soon as Possible." *Business Insider,* October
10, 2015, businessinsider.com/
elon-musk-mars-colonies-human-survival-2015-10?r=US&IR=T&IR=T.

Stephen Hawking on leaving Earth: M. Valle, "Slik tror Stephen Hawking
at vi kan forlate solsystemet" (This is how Stephen Hawking thinks we
can leave the solar system), *Teknisk Ukeblad,* June 21, 2017, tu.no/artikler/
slik-tror-stephen-hawking-at-vi-kan-forlate-solsystemet/396288.

Lecture with Kip Thorne: K. Thorne, "The Science of the Movie," lecture,
Realfagsbiblioteket, University of Oslo, September 7, 2016.

Interstellar: C. Nolan, dir., *Interstellar,* Legendary Pictures, 2014.

Closest neighboring star is four light years away: Wikipedia, "Proxima
Centauri," accessed February 21, 2020, en.wikipedia.org/wiki/
Proxima_Centauri.

10 | Can We Use Up the Earth?

Helium disappearing out into space: see heliumscarcity.com.

LIMITS FOR GROWTH

Limits to growth: D. H. Meadows et al., *The Limits to Growth (Where Is the
Limit?),* Universe Books, 1972.

Malthus warned in 1798: Jackson (2017); T. Malthus "An Essay on the
Principle of Population, as It Affects the Future Improvement of Society,
with Remarks on the Speculations of Mr. Goodwin, M. Condorcet, and
Other Writers," London: J. Johnson, 1798.

Smith in 1766: Raworth (2017), see page 250: "Adam Smith believed that
every economy would eventually reach what it called a 'stationary state'

with its 'full complement of riches' ultimately being determined by 'the nature of its soil, climate and situation'" (A. Smith, *An Inquiry into the Nature and Causes of the Wealth of Nations,* 1776)

Adam Smith was the founder of social economics: A. Sandmo, "Nasjonenes velstand" (The Wealth of Nations), Minerva, December 20, 2011, minervanett.no/nasjonenes-velstand/131848.

GROWTH THAT IS HAPPENING FASTER AND FASTER AND FASTER

The example of a magic drop of water in a sports arena is based on C. Martenson, "Crash Course Chapter 4: Compounding Is the Problem," Peak Prosperity (blog), peakprosperity.com/crashcourse/chapter-4-compounding-problem.

THE NECESSITY OF ECONOMIC GROWTH

The economy is built on growth, only going up or down, not straight ahead; economic growth without increasing resource use (decoupling) and the potential for economic stability without growth: See Raworth (2017) and Jackson (2017).

CAN THE ECONOMY GROW WITHOUT USING MORE RESOURCES?

More than 3 percent of electricity goes to the internet but could be 20 percent in five years: J. Vidal, "'Tsunami of Data' Could Consume One Fifth of Global Electricity by 2025," *Climate Home News,* December 11, 2017, climatechangenews.com/2017/12/11/tsunami-data-consume-one-fifth-global-electricity-2025.

THE HABITABLE ZONE

Revolt among economics students: See rethinkingeconomics.org.

Index

A

agriculture, 19–20, 132
  See also fertilizer; food production
Ajka, Hungary, 67
Alburnus Major, 31
  See also Roșia Montană
alcohol, in purification, 77–78
algae, 88, 147–48
alloys, 50, 70–71
aluminum, 66–70
aluminum oxide, 66, 67–68
ammonia, 120
antibiotics, 21
aquifers, 119
Årdal, Norway, 68–69
asteroids, 45, 155–58
atoms, 6

B

bacteria, 78, 79, 122
bacterial concrete, 97
Baekeland, Leo, 106
Bakelite, 106
batteries, 143, 144–45
bauxite, 66–67, 69
bedrock, 12, 26, 117
Bennu (asteroid), 157
big bang theory, 5–6
bioenergy, 146–48
bird dung, 123
Birkeland-Eyde process, 121, 122
Black Sea, 29
blood donation, 99–100
breathing, iron's role in, 44
burning used plastics, 110

C

Caesarea, Israel, 90–91
calcium hydroxide, 88–90, 91
carats, 36
carbon, 8, 45–46, 51–52, 101
    See also cellulose; lignin; plastics;
    rubber
carbon dioxide, in limestone, 88
cars, 61–62, 63, 71, 143–46
cast iron, 49–50, 71
celluloid, 104–5
cellulose, 103–5, 112
cement, 89–90, 91–92
ceramic materials
    about, 81–82
    clay, molding with, 83–84
    glass, 84–86
    living organisms as producers of,
        96–97
    properties, 82–83
    sand and gravel in, 94–96
    See also concrete
civilization, 19–21, 38–39, 132–33
clay, molding with, 83–84
cleanliness in making small
    machines, 77–78
climate change, 15, 17, 135
clovers, 120
coal, 46, 139
cobalt, 144, 145
communication systems, 86
computer-controlled cars, 62
concrete
    algae and, 88
    bacterial, 97
    Dubai, 95

history of, 88–91
    limestone, 97
    Minoans, 89–90
    reinforcing materials, 92–94
    Roman Empire, 90–91
    steel-reinforced, 92–94
    timber and, 91
    volcanic ash in, 90, 91
Congo Free State, 102
copper, 63–65
cryolite, 67–68, 69
cyanide, in gold extraction, 34–35
cyborgs, 76

D

Dacians, 30
dam failure, 67
Dead Sea, 114–15, 116, 119, 125,
    128
deforestation, 46, 64–65
degradable plastics, 111
deuterium, 152–53
dinosaurs, 17
distillation, for purification,
    77–78
Doughnut Economics (Raworth),
    171
Dubai, United Arab Emirates, 95
Dust Bowl, 125
dyes, creating solar cells with, 138

E

Earth
    energy stores, 131–32
    life on, 12–19, 159–60
    "using up," 161–62

economic growth
  limits to, 162–64
  necessity of, 165–68, 170–71
  resource limitations versus,
    168–69
Eiffel Tower, 50, 52–53
electric cars, 143–45
electricity, 62–63, 66, 75–76, 121,
  122, 169
electronics, 77–79, 86
electronic waste, gold in, 40
electrons, 6
energy
  about, 129–30
  bioenergy, 146–48
  for cars, 143–46
  civilization and, 132–33
  extracting, 133–34
  food production and, 132, 148
  fossil, 21, 132, 134–35, 148–49
  geothermal, 135–36
  hydroelectric, 138–39, 141–43
  nuclear, 136–37, 139
  solar, 13, 21, 130–31, 137–38, 141,
    148, 151–52
  stores, Earth's, 131–32
  storing, 142–43
  wind, 139–40, 141, 142–43
environmental effects
  aluminum production, 68–69
  copper mining and production,
    64–65
  gold mining, 32–34
  importance of, 2
  sand and gravel extraction, 95
  titanium mining and production,
    72–73

erosion of topsoil, 125–26
evaporation, 115, 119
exponential growth, 164–65
extinction, mass, 16–17

F
Fen Complex, 141
ferrous lava, 45–46
fertilizer, 113, 117, 121, 123, 127
  See also nitrogen; phosphorus;
    potassium
fire, 18, 30
flint, 87
fluoride, 67, 68, 69
fluorine, 68, 69
food packaging, 105
food production, 132, 148
forests, 46, 64–65, 91
fossil energy, 21, 132, 134–35, 148–49
fusion reactors, 151–54
future
  about, 151
  fusion reactors, 151–54
  humans and, 22–23
  planets, settlement on other,
    158–60
  space, elements in, 155–58

G
galaxies, 7
garbage, 108–9
Geamana, Romania, 33
genetically modified plants, 122
geothermal energy, 135–36
Germany, 48, 59
glass, 84–86

glazing ceramics, 84
global warming, 15, 17, 135
gold
  about, 25–26
  chemical reactions, 28
  civilization and, 38–39
  concentration of, 36
  extracting, 30–32
  first, 27
  lost, 39–41
  metal, making stone into, 34–35
  mining, environmental impacts
    of, 32–34
  recovering lost, 39
  resources, 39, 40–41
  in river gravel, 28–29
  Roșia Montană mines, 26, 28,
    29–32, 36, 37–38
  surface mining, 32
  volcanic activity and, 26–27
gold ore, 29–30, 33, 34, 35, 39, 41
grassy plains, 17
gravel, 28–29, 94–96
greenhouse effect, 15
Greenland, 45–46
groundwater, 34, 119
growth
  economic, necessity of, 165–68,
    170–71
  economic versus resource,
    168–69
  exponential, 164–65
  limits to, 162–64
  paradox of, 170
guano, 123
gutta-percha, 101

H
Haber-Bosch process, 121
Habsburgs, 31
hammered copper, 64
hand prosthesis, 75–76
health services, 99–100, 133
heavy water, 152
helium, 6, 7–8, 9, 161
Henderson Island, 107–9
Herod the Great (King of Judea),
  90–91
history of elements
  civilization, age of, 19–21
  humans and the future, 22–23
  life begins, 12–15
  living Earth, 16–19
  solar system, formation of our,
    5–12
  stars, life of, 7–9
  universe, birth of, 5–9
Hitler, Adolf, 48, 59
Hominoidea, 17
humans
  future and, 22–23
  as nomadic tribes, 18–19
  origin of, 18–19
  settlement of, 19–20
hydraulic cement, 89–90, 91–92
hydroelectric energy, 138–39,
  141–43
hydrogen, 6, 9
hydrogen cars, 145–46
hydrogen variants, 152

I

ice ages, 15, 16, 18
implants, 71–72
insect signaling systems, 76
internet, electrical energy used by, 169
Interstellar, 159
iron
    about, 43–44
    breathing, role in, 44
    carbon and, 45–46, 51–52
    cast, 49–50, 71
    coal and, 46
    Iron Age, 45–46
    metal, making ore into, 49–50
    meteorite, 45
    as nonrenewable resource, 161–62
    oxygen and, 44–45, 45–46
    pig, 49, 50, 56
    production/usage statistics, 54, 59
    resources, 54–59
    rust, problem with, 52–54
    scrap, 59
    steel, 50–52
    Swedish, 47–49, 56
    timber and, 46
    wrought, 50, 52
Iron Age, 45–46
iron carbide, 51
iron oxides, 46–47, 86
ironworks, 49–50
ITER (fusion experiment), 153

J

Jackson, Tim, 170–71
Jason and the Golden Fleece, 28–29
Jøssingfjord., 72–73

K

Kiruna, Sweden, 47–49, 56

L

Larsson, Arne, 75
lava, 45–46, 141
lead, 86, 137
LED bulbs, 168–69
Leopold II (King of Belgium), 102
lightning, 120–21, 122
lignin, 104, 112
limestone, 87–89, 90, 92
limestone concrete, 97
Limits to Growth, The, 162–63
lithium, 6, 143–44, 152–53
livestock, 20, 68
living organisms, as ceramic materials producers, 96–97
Luleå, Sweden, 47

M

magnetic field, 12
Malaysia, 67
mass, defined, 6
medical equipment, disposable, 99–100
mercury, 34
meteorite iron, 45
meteorites, 11, 155, 156
methane, 15

microorganisms, 106, 109, 111, 112

Milky Way, 7

miniaturization, 77–78

Minoans, 89–90

mobile phones, 70, 74

Morocco, 123–24

N

Narvik, Norway, 47, 49

natural gas, 121–22

Neanderthals, 18

neodymium, 140–41

neutrons, 6

neutron stars, 8–9

nitrate film, 104–5

nitrocellulose, 104–5

nitrogen, 119–22

　See also fertilizer

nomadic tribes, 18–19

Norsk Hydro, 121, 152

Norway, 72–73, 138, 139, 141, 148

nuclear energy, 136–37, 139

O

oil, 134

oil-based plastics, 106–7, 110, 111

organic molecules, 6, 15, 101, 122

outer space, elements in, 155–58

oxygen

　aluminum and, 66

　chemical composition, 8

　iron and, 44–45, 45–46

　solar system, formation of, 13–14,
　　15

ozone layer, 15

P

pacemakers, 75

packaging for food, 105

paint, 72

phosphorus, 122–26

　See also fertilizer

photosynthesis, 13–14, 130, 138, 148

pig iron, 49, 50, 56

planets, settling on other, 158–60

plasma, 153

plastics

　burning used, 110

　degradable, 111

　food packaging, 105

　non-oil-based, 112

　oil-based, 106–7, 110, 111

　recycling, 110

　in trash, 109

plate tectonics, 11–12

platinum, 146

polymers, 103, 110, 111, 112

population, human, 22, 131,
　162–65

Portland cement, 92

potassium, 116–17, 118–19

　See also fertilizer

Prosperity Without Growth (Jackson),
　170–71

protons, 6

R

railways, 43, 47–48

rare-earth elements, 140–41

Raworth, Kate, 171

recycling, 69–70, 87, 110, 140

river gravel, 28–29, 94–96

rocks, as ceramic material, 87
Roman Empire, 30–31, 32, 90–91
Romania, 35
Røros, Norway, 64–65
Roșia Montană, 26, 28, 29–32, 36, 37–38
rubber, 100–103
rust, 14, 46–47, 52–54, 93
Ryugu (asteroid), 157

S

sacrificial anodes, 53
sand, 94–96
scrap iron, 59
seabed, extracting phosphorus from, 124
self-driving cars, 62
Siberia, 16–17
Singapore, 95–96
slag, 49, 50
slaked lime, 88–90, 91
sludge, 33, 67, 72–73
solar cells, 137–38, 148
solar energy, 13, 21, 130–31, 137–38, 141, 148, 151–52
solar system, formation of our, 5–12
space, elements in, 155–58
space travel, 79
specialization, 132
stainless steel, 51, 53, 54, 59
stars, 7–9, 160
steel, 50–52, 92–94
stellarators, 154
stone, as ceramic material, 87
sulfur, 101–2, 103

supernovae, 8, 9
Swedish iron, 47–49, 56
synthetic plastics, 106–7, 110, 111

T

Telemark, Norway, 141
Tellnes titanium mine, 72–73
Thorne, Kip, 159
timber, 46, 64–65, 91
titanium, 71–74
titanium dioxide, 72
toilets, specialized, 126
tokamak, 153
tools, development of, 43–44, 132
topsoil, 125–26
trains, 43, 47–48
Transylvania, 25–26
   See also Roșia Montană
trash, 108–9
trees, 46, 64–65, 91, 103
tritium, 152
Tutankhamen (Egyptian pharaoh), 45, 155

U

Ulefoss, Norway, 141
universe, birth of, 5–9
urban mining, 40

V

Vikings, 49
viscose, 105
volcanic ash, 90, 91
volcanic eruptions, 16–17, 26–27, 85
vulcanization, rubber, 101–2, 103

W

water, 10–11, 119, 152
water pipes, copper, 63
wearable technology, 74
wedding rings, 25, 36
wind energy, 139–40, 141, 142–43
wind turbines, 139, 140, 141
World War II, 48, 59
wrought iron, 50, 52

About the Author

Anja Røyne, PhD, is a scientist and lecturer in the department of physics at the University of Oslo. A physicist with a background in solar energy, Røyne has also researched geological and geochemical processes and is now working on creating materials with biotechnology. In addition, she runs her own science blog, has shared her expertise in newspapers and radio programs, and frequently gives popular science talks.